Student Solutions Manual

Quantitative Reasoning
and the Environment
MATHEMATICAL MODELING IN CONTEXT

GREG LANGKAMP   JOSEPH HULL

Seattle Central Community College

PEARSON

Prentice Hall

Upper Saddle River, NJ 07458

Editor-in-Chief: Sally Yagan
Acquisitions Editor: Chuck Synovec
Project Manager: Michael Bell
Executive Managing Editor: Kathleen Schiaparelli
Senior Managing Editor: Nicole M. Jackson
Assistant Managing Editor: Karen Bosch Petrov
Production Editor: Jessica Barna
Supplement Cover Manager: Paul Gourhan
Supplement Cover Designer: Christopher Kossa
Manufacturing Buyer: Ilene Kahn
Manufacturing Manager: Alexis Heydt-Long

© 2007 Pearson Education, Inc.
Pearson Prentice Hall
Pearson Education, Inc.
Upper Saddle River, NJ 07458

Printed in the United States of America

10 9 8 7 6 5 4 3 2 1

ISBN 0-13-148531-8

Pearson Education Ltd., *London*
Pearson Education Australia Pty. Ltd., *Sydney*
Pearson Education Singapore, Pte. Ltd.
Pearson Education North Asia Ltd., *Hong Kong*
Pearson Education Canada, Inc., *Toronto*
Pearson Educación de Mexico, S.A. de C.V.
Pearson Education—Japan, *Tokyo*
Pearson Education Malaysia, Pte. Ltd.

Table of Contents

Chapter 1

1. Clam Shell

    Answers may vary. The shell is approximately 5.5 cm long and 4.1 cm wide. The shell is not a perfect rectangle, so the choice of axes may vary.

3. Reservoir Evaporation

    The actual rate of evaporation is 0.07 inches per day less than the measured value of 0.56 inches/day. The accuracy seems moderate but not too high.

5. MTBE in Groundwater.

    The precision seems very high. The scientists were able to measure to within one part out of a billion parts, which is a very tiny fraction.

7. Gallons of Gasoline

    Answers may vary. An average person might drive about 10,000 miles in one year. A typical automobile fuel economy might be 20 miles per gallon of gasoline. If 1 gallon of gasoline yields 20 miles of driving, then 500 gallons yields 10,000 miles of driving.

9. Food Consumption

    Answers may vary. If an individual eats 1 pound of food per meal on average, then that person eats 3 pounds per day. Over a year's time (365 days), that individual will consume about 1100 pounds of food (over half a ton!). One pound per meal may be excessive; we normally think about food in terms of calories (energy), not weight. This estimate neglects snacking.

11. A metric unit of mass is the kilogram and a metric unit of energy is the joule.

13. "Acre-foot" is a measure of volume (one acre multiplied by one foot).

15. Forest Transect

    Feet must cancel to leave miles.

    $$13{,}267 \text{ feet} \times \frac{1 \text{ mile}}{5{,}280 \text{ feet}} = 13{,}267 \text{ feet} \times \frac{1 \text{ mile}}{5{,}280 \text{ feet}} \approx 2.51 \text{ miles}$$

17. Wetlands

Square kilometers must cancel to leave hectares.

$$1.45 \ \cancel{km^2} \times \frac{100 \, ha}{1 \ \cancel{km^2}} = 145 \, ha$$

19. Pickup Power

The units of horsepower must cancel to leave kilowatts.

$$137 \ \cancel{horsepower} \times \frac{0.7457 \, kilowatts}{1 \ \cancel{horsepower}} \approx 102.16 \, kW$$

$$133 \ \cancel{hp} \times \frac{0.7457 \, kW}{1 \ \cancel{hp}} \approx 99.18 \, kW$$

21. Earth's Circumference

The number of miles must be less than the number of kilometers.

$$40,000 \ \cancel{km} \times \frac{1 \, mile}{1.609 \ \cancel{km}} \approx 24,860 \, miles$$

23. Discharge

Break this calculation into two stages. For example, convert minutes to hours first, then convert gallons to cubic meters.

$$\frac{3.5 \, gallons}{\cancel{minute}} \times \frac{60 \ \cancel{minutes}}{1 \, hour} = \frac{210 \, gallons}{1 \, hour}$$

$$\frac{210 \ \cancel{gallons}}{1 \, hour} \times \frac{1 \, m^3}{264.172 \ \cancel{gallons}} = 0.795 \frac{m^3}{hour}$$

25. River Sediment

The sediment accumulated in a 90-year period from 1913 to 2002.

$$\frac{2.4 \text{ million cubic yards}}{90 \text{ years}} = \frac{26,667 \, yds^3}{1 \, year}$$

$$\frac{26{,}667\,\text{yds}^3}{1\,\text{year}} \times \frac{1\,\text{year}}{365\,\text{days}} = 73.1\frac{\text{yds}^3}{\text{day}}$$

27. Distance and Time

   The units are meters per second $\left(\dfrac{m}{sec}\right)$.

29. The abbreviation for a million watts of electricity is a megawatt (MW).

31. A μm is one millionth (10^{-6}) of a meter in distance.

33. Mammals Versus Insects

   There were 4.65×10^3 named species of mammals and 1.025×10^6 named species of insects, such that there were about 200 times as many insects as mammals ($\dfrac{1.025\times10^6}{4.65\times10^3}\approx2\times10^2$). These two values therefore differ by about 2-3 orders of magnitude (between 100 and 1,000 times).

35. PM10

   Chicago's air had $0.000025\dfrac{g}{m^3}$ or $2.5\times10^{-5}\dfrac{g}{m^3}$ of particulates, whereas the EPA standard is $0.00015\dfrac{g}{m^3}$ or $1.5\times10^{-4}\dfrac{g}{m^3}$. These two values differ by about one order of magnitude.

37. The log of 10^{-6} is -6, the log of 10^8 is 8, the log of 100,000 is 5 and the log of 0.001 is -3.

39. The log of 1,731,124 is 6.238 and the log of 0.000352 is -3.453. 1,731,124 is $10^{6.238}$ and 0.000352 is $10^{-3.453}$.

41. Wildfire Acreages

The logarithm of 101,013 acres is 5.004 log acres and the log of 1,676,414 acres is 6.224 log acres. These two acreages differ by about 1 order of magnitude $(6.2-5.0 \approx 1)$.

43. Indonesian Earthquake

Substitute the size of the earthquake in the equation and solve for the magnitude. Use extra parentheses on your calculator as needed.

$$\text{magnitude} = \log\left(\left(3.5 \times 10^{29} \ \text{dyne} \times \text{cm}\right)^{\frac{2}{3}} \Big/ 10^{10.7}\right) \approx 9.0$$

45. Mine drainage

$\text{pH} = -\log\left(1.6 \times 10^{-5} \ \text{moles/liter}\right) \approx 4.8$. The lake water is rather acidic.

Chapter 2

1. Red List

$$\frac{15{,}042 \text{ species}}{10{,}731 \text{ species}} \approx 1.40$$

3. Land Areas

Set up the ratio equation, cross multiply, and solve for California's area.

$$\frac{4 \text{ million km}^2}{\text{California's area}} = 10$$

$$\text{California's area} = \frac{4 \text{ million km}^2}{10} = 400{,}000 \text{ km}^2$$

5. California Power Plants

   a. Divide carbon dioxide output by energy input.

Plant Name	energy input (TBTU)	CO_2 output (tons)	normalized (tons/TBTU)
Santa Clara	833,261	48,213	0.058
SCA	4,487,554	2,340,088	0.521
Scattergood	9,889,331	537,922	0.054
South Bay	23,038,851	1,353,523	0.059
SPA	2,460,438	145,701	0.059
Walnut	40,594	2,349	0.058

   b. With one exception (SCA), all the power plants have very similar ratios of carbon dioxide output to energy input, suggesting they are of similar design and burning similar fuels, with similar efficiencies.

   c. SCA put out the most greenhouse gas per energy input, about half a ton of carbon dioxide for each trillion BTU.

7. San Francisco Automobiles
   a. The units are cars/person.

Year	Cars	Population	Normalized (cars/person)
1930	461,800	1,578,000	0.293
1940	612,500	1,734,300	0.353
1950	1,006,400	2,681,300	0.375
1960	1,620,600	3,638,900	0.445
1970	2,503,100	4,630,600	0.541
1980	3,281,800	5,179,800	0.634
1990	3,953,200	6,023,600	0.656
2000	4,799,300	6,875,400	0.698

   b. Per capita car ownership has increased every year since 1930, but seems to have increased more slowly over the last few decades.

9. Forest Areas
   a. Divide the area protected by the total forest area and multiply by 100%.

Country	Forest area (km^2)	Protected Forest (km^2)	Percent Protected
Cameroon	289,965	17,854	6.2%
Central African Republic	199,018	41,608	20.9%
Congo	278,797	12,935	4.6%
Equatorial Guinea	23,540	0	0.0%
Gabon	239,369	8,975	3.7%
Zaire	1,439,178	93,160	6.5%

   b. Zaire, a very large country, has the most protected forest.
   c. The Central African Republic has the greatest percentage of protected forest.

11. Water Bottles.
    If 16% of the bottles are recycled, then 84% (0.84) are discarded.

    0.84 × 1 billion bottles/year = 840 million bottles/year discarded

13. U.S. Birth Rate

There are several approaches to this exercise. Here we calculate the ratio of births to people, and then multiply by $1,000/1,000 \, (=1)$.

$$\frac{4,137,000 \text{ births}}{281,422,000 \text{ people}} \approx 0.0147 \times \frac{1,000}{1,000} = \frac{14.7}{1000} = 14.7 \, \text{ppt} = 14.7 \, {}^{0}/_{00}$$

15. Lead in Trout

Express the ratio in terms of like units; for example, grams. Then simplify.

$$\frac{1 \text{ milligram}}{1 \text{ kilogram}} = \frac{10^{-3} \text{ gram}}{10^3 \text{ gram}} = \frac{1}{10^3 \times 10^3} = \frac{1}{1 \text{ million}} = \frac{1 \text{ part}}{1 \text{ million parts}} = 1 \, \text{ppm}$$

One milligram per kilogram is the same as 1 part per million.

$$\frac{1}{1 \text{ million}} \times \frac{1,000}{1,000} = \frac{1,000}{1 \text{ billion}} = 1,000 \, \text{ppb}$$

One milligram per kilogram is also the same as 1,000 parts per billion.

17. Prairie Dogs

Use the percentage change formula; the units "prairie dogs" will cancel.

$$\frac{30 \text{ dogs} - 10 \text{ dogs}}{10 \text{ dogs}} \times 100\% = \frac{20}{10} \times 100\% = 200\%$$

$$\frac{300 \text{ dogs} - 100 \text{ dogs}}{100 \text{ dogs}} \times 100\% = \frac{200}{100} \times 100\% = 200\%$$

The two colonies show the same percentage change with time; both increased by 200%.

19. Toxics

Use the percentage change formula.

$$\frac{7.1 \text{ billion lbs} - 7.7 \text{ billion lbs}}{7.7 \text{ billion lbs}} \times 100\% = -7.79\%$$

The amount of toxics decreased by about 8% from 1999 to 2000. The sign is negative, consistent with a decrease in reported toxic wastes.

21. England and Wales Air Pollution

The unknown in the percentage change formula is the initial number of air pollution incidents in year 2001. We note that a 54% decrease is equivalent to -0.54:

$$\frac{225 \text{ incidents} - \text{initial}}{\text{initial}} = -0.54$$

Let x = the initial amount:

$$\frac{225 - x}{x} = -0.54$$

Solve for x by first cross-multiplying, and then adding together like terms:

$$225 - x = -0.54x$$
$$225 = -0.54x + x$$
$$225 = 0.46x$$
$$489 \approx x$$

There were approximately 489 air pollution incidents in 2001.

23. Idaho Wheat Farms

Use the percentage difference formula, with District 1 (1,259 acres) as the reference value.

$$\frac{789 \text{ acres} - 1{,}259 \text{ acres}}{1{,}259 \text{ acres}} \times 100\% = -37.3\%$$

Wheat farms in District 3 are approximately 37% smaller in average acreage than those in District 1.

$$\frac{1{,}690 \text{ acres} - 1{,}259 \text{ acres}}{1{,}259 \text{ acres}} \times 100\% = 34.2\%$$

Wheat farms in District 5 are approximately 34% larger in average acreage than wheat farms in District 1.

25. Laboratory Analysis

Use the percentage error formula, with 150 ppb as the known or correct value.

$$\frac{159 \text{ ppb} - 150 \text{ ppb}}{150 \text{ ppb}} \times 100\% = 6.0\%$$

27. Daily Water Use

Let x be the city's daily water use. Multiply both sides by 115,000 people. The units will cancel, leaving "gallons."

$$\frac{50 \text{ gallons}}{1 \text{ person}} = \frac{x}{115{,}000 \text{ people}}$$

$$x = \frac{\left(50 \text{ gallons}\right)\left(115{,}000 \text{ people}\right)}{1 \text{ person}} = 5.75 \text{ million gallons}$$

29. Areas

Let x be the number of hectares in 1 square mile. Do not include units for x; the units will cancel appropriately.

$$\frac{100 \text{ ha}}{0.386 \text{ mi}^2} = \frac{x}{1 \text{ mi}^2}$$

$$x = \frac{100\,\text{ha} \times 1\ \text{mi}^2}{0.386\,\text{mi}^2} \approx 259.07\,\text{ha}$$

31. Brownfield Reclamation

Let x be the percentage of reclaimed acreage. 28.7 acres is 100% of the area, therefore:

$$\frac{3.2\ \text{acres}}{28.7\ \text{acres}} = \frac{x}{100\%}$$

$$x = \frac{3.2\ \text{acres} \times 100\%}{28.7\ \text{acres}} \approx 11.1\%$$

33. PBDE in Fish

Let x be the unknown number of parts in a billion parts:

$$\frac{3,078}{1\ \text{trillion}} = \frac{x}{1\ \text{billion}}$$

$$x = \frac{3,078 \times 1\ \text{billion}}{1\,\text{trillion}} = \frac{3,078 \times 10^9}{10^{12}} = 3,078 \times 10^{-3} = 3.078$$

The concentration of PBDE in the fish sample is 3.078 ppb.

35. Desert Survey

a. Let x be the density of palo verde in the second area:

$$\frac{270\ \text{palo verde/ha}}{315\,\text{ironwood/ha}} = \frac{x}{120\,\text{ironwood/ha}}$$

$$x = \frac{120\,\text{ironwood/ha} \times 270\ \text{palo verde/ha}}{315\,\text{ironwood/ha}} \approx 103\,\text{palo verde/ha}$$

b. Let y be the density of saguaro in the second area:

$$\frac{165 \text{ saguaro/ha}}{315 \text{ ironwood/ha}} = \frac{y}{120 \text{ ironwood/ha}}$$

$$y = \frac{120 \text{ ironwood/ha} \times 165 \text{ saguaro/ha}}{315 \text{ ironwood/ha}} \approx 63 \text{ saguaro/ha}$$

37. Canyon Treefrog

A quick sketch might help to visualize the components of the capture-recapture method. Use the capture-recapture formula to set up the proportion and then solve for N, the size of the population.

$$\frac{61 \text{ marked tadpoles in population}}{36 \text{ marked tadpoles in recapture}} = \frac{N \left(\text{tadpoles in population} \right)}{68 \text{ tadpoles in recapture}}$$

$$\frac{\left(61 \text{ marked tadpoles in population} \right) \left(68 \text{ tadpoles in recapture} \right)}{36 \text{ marked tadpoles in recapture}} = N \approx 115 \text{ tadpoles}$$

Boot Spring was probably a closed habitat. The tadpoles were recaptured after a short period of time, only 2 hours. The tadpoles probably didn't travel that far in 2 hours. Therefore the study matched the assumptions behind the catch and release method.

39. New York Pickerel

Use the capture-recapture formula to set up the proportionality.

$$\frac{232 \text{ marked in population}}{16 \text{ marked in sample}} = \frac{N}{329 \text{ pickerel total in sample}}$$

$$N \approx 4,771 \text{ pickerel}$$

Most lakes are relatively closed environments for most fish, especially pickerel, which are not migratory like salmon. The closed system will help the quality of the population estimate; large net migrations into or out of the ecosystem between the first and second sampling would alter the size of the population. It's harder to tag a moving target.

41. Tumbling Dice

   a. There are a total of 36 possible outcomes:

1-1	2-1	3-1	4-1	5-1	6-1
1-2	2-2	3-2	4-2	5-2	6-2
1-3	2-3	3-3	4-3	5-3	6-3
1-4	2-4	3-4	4-4	5-4	6-4
1-5	2-5	3-5	4-5	5-5	6-5
1-6	2-6	3-6	4-6	5-6	6-6

   b. Out of 36 possible outcomes, there are 6 successful outcomes (sum = 7):

1-1	2-1	3-1	4-1	5-1	**6-1**
1-2	2-2	3-2	4-2	**5-2**	6-2
1-3	2-3	3-3	**4-3**	5-3	6-3
1-4	2-4	**3-4**	4-4	5-4	6-4
1-5	**2-5**	3-5	4-5	5-5	6-5
1-6	2-6	3-6	4-6	5-6	6-6

   The probability of rolling a sum equal to 7 is therefore:

$$P(\text{sum}=7)=\frac{6}{36}=0.1\overline{6}\approx 17\%$$

   c. Out of 36 possible outcomes, there are 4 successful outcomes (sum = 9). The probability of rolling a sum equal to 9 is:

$$P(\text{sum}=9)=\frac{4}{36}=0.\overline{1}\approx 11\%$$

   d. There are 21 combinations whose sum is greater than 6. The probability of rolling one of these combinations is:

$$P(\text{sum} > 6) = \frac{21}{36} = 0.58\overline{3} \approx 58\%$$

e. There are 4 combinations whose sum is 5. The probability of rolling 2 dice whose sum is *not* 5 is:

$$P(\text{sum} \neq 5) = 1 - P(\text{sum} = 5) = 1 - \frac{4}{36} = 0.\overline{8} \approx 89\%$$

43. Mississippi River Floods
   a. The probability of a flood in any year is:

$$P(\text{flood}) = \frac{26 \text{ years with floods}}{99 \text{ years possible}} = 0.\overline{26} \approx 26\%$$

   b. The probability of a flood *not* occurring in any year is:

$$P(\text{no flood}) = 1 - P(\text{flood}) = 1 - 0.\overline{26} = 0.\overline{73} \approx 74\%$$

45. Cascadia Earthquakes
   a. The average recurrence interval is:

$$R = \frac{3{,}200 \text{ years}}{6 \text{ intervals}} \approx 533 \,{}^{\text{years}}\!/_{\text{interval}}$$

   b. We should not expect another great earthquake 200 years from now, as the actual recurrence intervals between each of those 7 great earthquakes probably varied quite a bit. The next great earthquake may be sooner or later than 200 years.

47. Colorado Wildfires
   There were 36 large wildfires started in the 135 days between April 23 (day 113) and September 5 (day 248).

$$R = \frac{135 \text{ days}}{35 \text{ intervals}} \approx 3.9 \text{ days/interval}$$

Chapter 3

1. Greenhouse Gases.

   Estimates may vary. The actual values are shown below.

Gas	CO_2	CH_4	N_2O	Total
%	66%	18%	16%	100%

3. Toxic Release Inventory

   a. Start by calculating the total toxics released, then calculate the corresponding percentage for each state. Check your mathematics by summing the percentages. In the pie chart, include a state identification.

State	Toxics	Percentage
Nevada	1,000	28.3%
Utah	956	27.0%
Arizona	744	21.0%
Alaska	535	15.1%
Texas	302	8.5%
Total	3,537	99.9%

   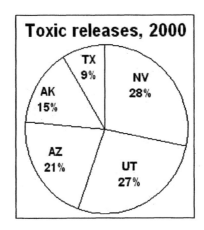

   b. Nevada and Utah release approximately twice as much toxic material as Alaska.
   c. The values in the table (1 billion pounds versus 535 million pounds) support this assertion.

14

5. Recycled Materials
   a. Begin by calculating the total recycling for 1986 and 1998, then use these totals to calculate the percentages for each category. The percentages should sum to 100%, with a bit of round-off error.

Category	1986	1986 %	1998	1998 %
Papers	391,994	87.1%	821,994	37.9%
Metals	9,528	2.1%	318,710	14.7%
Organics	0	0.0%	815,809	37.6%
Plastics	349	0.1%	9,871	0.5%
Glass	48,013	10.7%	113,338	5.2%
Others	352	0.1%	87,657	4.0%
Totals	**450,236**	**100.1%**	**2,167,379**	**99.9%**

   b. Organics, plastics, metals and others are not apparent on the 1986 chart. All recyclables are shown on the 1998 chart.
   c. Organics, plastics and others have increased as percentages of the total from 1986 to 1998. Paper and glass have declined as a percentage of the total.

7. Ecological Footprint
   a. The units for the ecological footprint are hectares per person or hectares per capita.
   b. Renewable resources would include grains like rice and wheat, trees for lumber, fruit-bearing trees, and grass and forage produced on grazing lands.
   c. The ratio of the Canadian footprint to the Haitian footprint is about 8.8 ha/person to 0.8 ha/ person or 11 to 1.

9. U.S. Energy Consumption
    a. *Results my vary.* In bar charts, the bars are typically separated by gaps. In this
       example, the bars can be placed along the horizontal axis in any order.

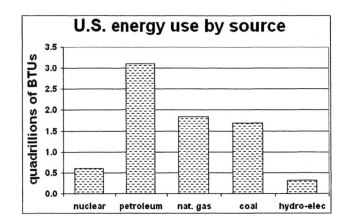

    b. Only hydroelectricity is renewable, as rainfall restores the water behind dams.
       The others are finite resources.
    c. Solar, geothermal, and wind energy (all renewable) are some of the sources that
       are missing from the bar chart.

11. Bangladesh Females
    a. The bin width is 5 years.
    b. The 80+ bin is the only bin that is not 5 years wide.
    c. *Results may vary.* Summing the heights of the bars, we estimate that there were
       approximately 63 million females in Bangladesh in 2000.

13. Mexico Population Pyramid
    a. The bin width is 5 years.
    b. All of the bins have the same width except bin 80+, which contains individuals
       equal to or older than 80 years old.
    c. Females begin to outnumber males in the 25-30 age interval.
    d. *Results may vary.* There were approximately 4 to 4.5 million people in their
       sixties in Mexico in 2000.

15. Tennessee Annual Precipitation
    a. *Results may vary* depending upon the choice of bins. We chose a bin width of 5
       inches, starting at 25 inches. Summing the frequencies of all the bins provides a

check on the mathematics. This diagram is a histogram, not a bar chart, therefore the bins touch.

inches	frequency
25-30	1
30-35	0
35-40	5
40-45	11
45-50	10
50-55	8
55-60	3
60-65	3
65-70	2
70-75	1
Total	**44**

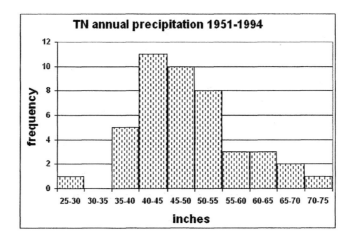

b. There is one empty bin from 30-35 inches of precipitation per year.
c. The modal bin is 40-45 inches of precipitation per year.
d. Six values fall below the modal bin and 27 values exceed the modal bin.

17. Annual Precipitation in Pennsylvania
   a. Sum the frequencies, then calculate the relative frequencies by dividing each frequency by the total number of measurements.

Reading			Montrose		
inches/year	frequency	rel.freq.	inches/year	frequency	rel. freq.
25-30	2	0.015	25-30	0	0.000
30-35	19	0.144	30-35	6	0.171
35-40	27	0.205	35-40	6	0.171
40-45	39	0.295	40-45	9	0.257
45-50	33	0.250	45-50	10	0.286
50-55	10	0.076	50-55	2	0.057
55-60	1	0.008	55-60	2	0.057
60-65	0	0.000	60-65	0	0.000
65-70	1	0.008	65-70	0	0.000
Total	132	1.001	Total	35	0.999

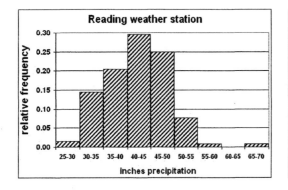

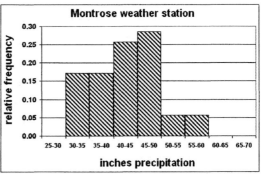

b. The modal bin for Reading is 40-45 inches of precipitation per year, whereas the modal bin for Montrose is 45-50 inches.

c. The relative frequencies for the two modal bins are about the same (0.29 and 0.28).

d. Reading has the longer period of record, and does have more extreme drought and deluge than Montrose. Reading has had an annual precipitation of 25-30 inches and 65-70 inches, whereas Montrose has not had such extremes.

19. Wisconsin Gray Wolves

a. The two variables are time (in years) and number of wolves.

b. *Answers may vary.* The minimum number of wolves was about 15 and the maximum number of wolves was about 330.

c. The wolf population took off in 1994. There are many possible reasons for this dramatic increase. Hunting and trapping of wolves may have been banned in Wisconsin around 1994. The deer population might have increased, providing more food. Or more wolves may have been introduced to Wisconsin from other places.

21. Recycling Aluminum

a.

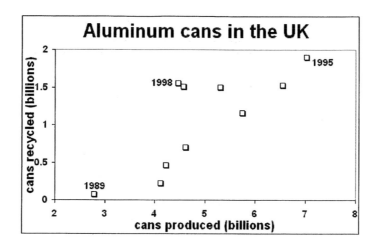

b. In general, the higher the amount of cans produced, the higher the number of cans recycled.

23. U.K. Aluminum Cans

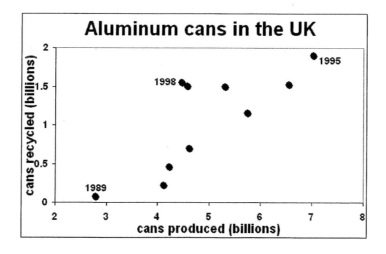

Chapter 4

1. Writing linear functions

    a. Let GT equal global temperature in °F, and t the time in years since 2001.

       $GT = 58.2 + 0.03t$.

    b. Let F equal the U.S. total fertility rate, and t the years since 2001.

       $F = 2.034 - 0.022t$.

3. Two tables

    a. Example A: $m = 400$ people/year   Example B: $m = -550 \dfrac{\text{people/km}^2}{\text{km}}$

    b. Example A: $P = 2,000 + 400t$ Example B: $P = 5,000 - 550d$

    c.

Example A		Example B	
t	P	d	P
0	2,000	0	5,000
1	2,400	1	4,450
2	2,800	2	3,900
3	3,200	3	3,350
4	3,600	4	2,800

5. Two tables

    a. Example A: $m = 750 \dfrac{\text{kWh}}{\$1,000}$ Example B: $m = -55 \dfrac{\text{million gallons}}{\text{year}}$

    b. Example A: $E = 10,500 + 750I$ Example B: $C = 2,925 - 55t$

    c.

Example A		Example B	
I	E	t	C
0	10,500	0	2,925
2	12,000	5	2,650
4	13,500	10	2,375
6	15,000	15	2,100
8	16,500	20	1,825

7. Balancing units

The right side of the equation reduces to units of millions of hectares when years are cancelled; those are the same units as on the left side of the equation.

$$F(\text{Mha}) = 3,510(\text{Mha}) - 11.2\left(\frac{\text{Mha}}{\cancel{\text{yr}}}\right) \times t(\cancel{\text{yr}})$$

$$\text{Mha} = \text{Mha}$$

9. U.S. population

a. Let P represent the U.S. population in millions of people and t the number of months since April 1, 2000.

b. Slope $= \dfrac{P_2 - P_1}{t_2 - t_1} = \dfrac{284.8 - 281.4 \text{ million people}}{15 - 0 \text{ months}} = 0.2267$ million people/month

The slope indicates that the U.S. was gaining, on average, 226,700 people per month during this time period.

c. $P = 281.4 + 0.2267t$

d. Solve $300 = 281.4 + 0.2267t$ to get $t \approx 82$ months. This is 6 years and 10 months from April 1, 2000, or February 1, 2007.

e. See graph. Extrapolation to 240 months is a bit extreme.

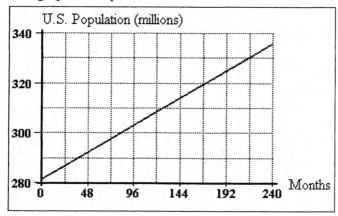

11. Soil thickness in northern Midwest states

a. Using the (t, S) points $(0, 0)$ and $(10000, 30)$, the slope is equal to $m = 0.003$.
Because $b = 0$, the equation is $S = 0.003t$.

b. The slope tells us the rate that soil forms: 0.003 in/yr. That's slow!

c.

Years	Inches	Years	Inches
0	0	6,000	18
1,000	3	7,000	21
2,000	6	8,000	24
3,000	9	9,000	27
4,000	12	10,000	30
5,000	15		

d. In the equation, let soil thickness equal 1/8 or 0.125. Then solve the equation $0.125 = 0.003t$ to get $t \approx 42$ years .

13. Energy consumption as a function of wealth

   a. Let E represent energy consumption (quadrillion BTUs) and GDP represent Gross Domestic Product (billion U.S.$). Spain and the U.S. are represented by the (GDP, E) data points (723, 5.7) and (9040, 97.1), respectively. The slope of the line between these points is about $m = 0.011$. The equation of the line that approximates these points is $E = 0.011GDP - 2.25$.

   b.

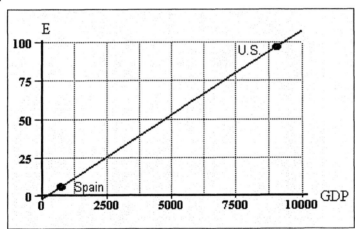

   c. Units of slope are $\dfrac{\text{quadrillion BTUs}}{\text{billion U.S.\$}}$. The slope indicates that energy consumption climbs 0.011 quadrillion BTUs (11 trillion BTUs) for every 1 billion dollar increase in GDP.

d. Spain: $\dfrac{5.7 \text{ quadrillion BTUs}}{723 \text{ billion \$}} = 7{,}884$ BTU/\$.

U.S.: $\dfrac{97.1 \text{ quadrillion BTUs}}{9{,}040 \text{ billion \$}} = 10{,}741$ BTU/\$. The U.S. uses more energy.

15. Coal on planet Earth

a. Consumption rate: $9.83 \times 10^{16}\ \dfrac{\text{BTU}}{\text{yr}} \times \dfrac{1 \text{ ton}}{2.068 \times 10^{7} \text{ BTU}} \approx 4.8 \times 10^{9}\ \dfrac{\text{ton}}{\text{yr}}$ or 4.8 billion tons per year.

b. Equation: $A = 1{,}081 - 4.8t$.

c. In year 2052, when $t = 50$, the world's supply of coal would equal $A = 1{,}081 - 4.8(50) = 841$ billion tons.

d. Substitute 200 for A and solve the resulting equation: $200 = 1{,}081 - 4.8t$. The answer is $t \approx 184$ years after 2002, or the year 2186.

e. Annual world coal consumption has risen tremendously because of the increase in the world's population, and because people are using more energy today. One might expect the consumption rate to climb well above 4.8 billion tons per year.

17. Water quality at a Florida chicken farm

a. See the following graph.

b. Most of the data look quite linear, except for the one significant outlier: (0.6, 26)

c. *Solutions may vary.* The next graph shows the data (solid circles) and a line drawn through the data using the straightedge method. The points chosen (open circles) to find the equation of the line are (0, 2) and (1.8, 24). The slope through these points is $m \approx 12.2$. Let N represent the nitrate concentration and P the potassium concentration, both in milligrams per liter. The equation of line is $N = 12.2P + 2$.

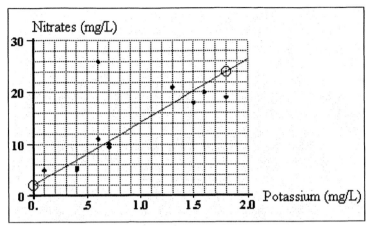

d. The calculator's regression line is $N = 9.517P + 4.8981$. The correlation coefficient is $r = 0.71$; this value tells us that potassium concentration is a fairly good predictor of nitrate concentration.

e. The nitrate concentration would equal approximately $N = 15.4$ mg/L.

19. Hawaii-Emperor chain

a. See next graph.

b. The data are very linear, except for the point corresponding to the island of Kimmei. It appears that Kimmei's estimated age is too low.

c. *Solutions may vary.* The next graph shows the data (solid circles) and a line drawn using the straightedge method. The points chosen (open circles) to find the equation of the line are (10, 1000) and (50, 4000). Let D represent distance in kilometers, and a the age in millions of years. The equation of the line is $D = 75a + 250$.

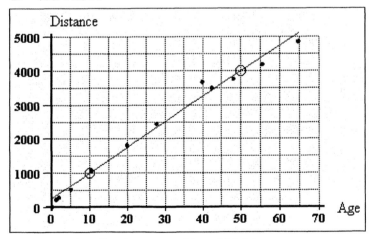

d. The regression line is $D = 75.75a + 184.84$.

e. The correlation coefficient is $r = 0.9944$. The size of the correlation coefficient indicates that the data are very linear, since the value is close to 1.0. The sign of the correlation coefficient indicates that the regression line has a positive slope.

f. Use the regression equation to set up the equation $1,841 = 75.75a + 184.84$. Solve to get the age: $a = 21.86$ million years old.

21. Columbia River velocities

a. velocity $= -0.1164 \, \text{depth} + 1.6801$

b. Depth is a good predictor of velocity. We can tell this by the correlation value of $r = -0.94$, which is very close to -1.0

c. The slope of the regression line tells us how rapidly the velocity decreases as depth increases; for every 1 foot increase in depth, the velocity decreases by 0.1164 feet per second.

d. The horizontal intercept can be found by setting velocity equal to 0, and solving for depth. We get 14.4 ft. This represents the hypothetical depth at which velocity diminishes to 0.

e. The closer to the bottom, the greater the frictional forces or "drag" on the water caused by rocks, vegetation, debris, etc.

Chapter 5

1. Writing exponential functions

    a. Let t represent years since 2000, and P the population in millions. $P = 99.9(1.012)^t$

    b. Let t represent years after 1988, and P the world production of CFCs in metric tons.

      $P = 1,074,465(0.734)^t$

3. Two tables

    a. Example A: $M = 1.2$ Example B: $M = 0.95$

    b. Example A: $r = 0.2 = 20\%$ Example B: $r = -0.05 = -5\%$

    c. Example A: $A = 20(1.2)^t$ Example B: $P = 120,000(0.95)^t$

Example A	
t	A
0	20
1	24
2	28.8
3	34.56
4	41.472

Example B	
t	P
0	120,000
1	114,000
2	108,300
3	102,885
4	$\approx 97,741$

5. Solving exponential equations

    a. $x = \dfrac{\log(40)}{\log(4)} \approx 2.661$ b. $W = \dfrac{\log(10)}{\log(1.0625)} \approx 37.981$

    c. $t = \dfrac{\log(150/52)}{\log(3)} \approx 0.964$ d. $x = \dfrac{\log(0.06)}{\log(4)} \approx -2.029$

7. Biodiesel

    a. Let B represent biodiesel production in millions of gallons, and t the number of

      years after 1999. $B = 5(1.431)^t$

b.

t	B
0	5
1	7.2
2	10.2
3	14.7
4	21.0
5	30.0

t	B
6	42.9
7	61.4
8	87.9
9	125.8
10	180.0
11	257.6

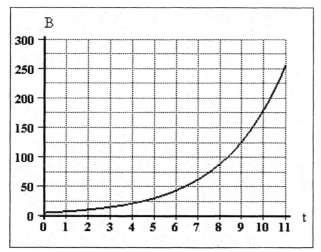

c. Yes, biodiesel production increased six-fold, from 5 million gallons to 30 million gallons in the period 1999-2004. It will take another five years for biodiesel to increase by six-fold again (the period 2004-2009).

9. Shrinking populations

a. Belarus: $P(t) = 10.4(0.9983)^t$; Hungary: $P(t) = 10.1(0.9971)^t$;

b. Hungary has the smaller multiplier, so its population is declining the fastest.

c. In 2015 ($t = 15$), Belarus will have 10.14 million people while Hungary will have 9.67 million people. That's a difference of about 0.47 million or 470,000 people.

11. Vehicle miles in U.S.

a. The number of vehicle miles in 1988 was 1,511 billion and six years later it was 1,793 billion. Solve the equation $1,511M^6 = 1,793$ to get $M = (1,793/1,511)^{1/6} \approx$ 1.0289. The yearly growth rate is $r = 0.0289$ or 2.89%.

b. Let *VM* represent vehicle miles in billions, and t the number of years since 1988.

$VM = 1,511(1.0289)^t$.

c. In the year 2040, $t = 52$. The number of vehicle miles will approximately equal 6,648 billion miles.

d.

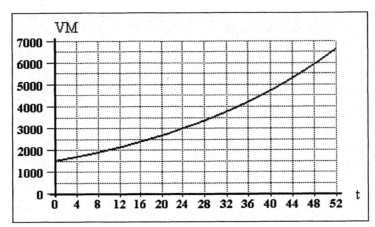

e. The graph indicates that vehicle miles doubles from about 1,500 billion miles when $t = 0$ to 3,000 billion miles when $t = 24$. The miles double from 3,000 to 6,000 billion miles between years $t = 24$ and $t = 48$. It appears that the doubling time is approximately 24 years. To find the doubling time more accurately, solve the equation $1,511(1.0289)^t = 3022$ which is equivalent to solving the equation

$1.0289^t = 2$. The solution is $t = \dfrac{\log(2)}{\log(1.0289)} \approx 24.3$ years.

f. Vehicle miles will most likely grow exponentially because the U.S. population is growing exponentially, and limitations in space seem to be offset by people's increasing desire to travel in vehicles.

13. Timber harvests in California

a. To find the yearly decay multiplier, solve the equation $2.27M^6 = 1.69$. The result is $M = (1.69/2.27)^{1/6} \approx 0.9520$. Let t represent the number of years since 1996, and H the harvest in billions of board feet. $H = 2.27(0.9520)^t$.

b. Using the equation $M = r + 1$, we find that the annual decay rate is $r = -0.048$ or -4.8% each year.

c. Find the timber harvest in 1998 by letting $t = 2$: $H = 2.27(0.9520)^2 \approx 2.06 \times 10^9$

board feet. The number of houses that could be built was: $\dfrac{2.06 \times 10^9}{14,000} \approx 147,000$.

15. Genetic diversity of domestic livestock

    a. $DLB = 5,000(0.95)^t$ where DLB represents the number of domestic livestock breeds and t the years, with $t = 0$ referring to the year when $DLB = 5,000$.

    b. Solve the equation $2,500 = 5,000(0.95)^t$. First divide by 5,000; then take the log of each side. Isolate t to get $t = \log(0.5)/\log(0.95) \approx 13.5$ years. Interpretation: Every 13.5 years, the number of domestic livestock breeds will be reduced by 50%.

17. The three fastest growing U.S. counties

    a. If M equals the monthly multiplier, then M^{15} will equal the 15-month multiplier. Douglas County, CO: solve $M^{15} = 1.136$ to get $M = 1.136^{1/15} \approx 1.00854$. This makes the monthly growth rate approximately 0.00854 or 0.854%.

    Loudoun County, VA: solve $M^{15} = 1.126$ to get $M \approx 1.00794$. The monthly growth rate is approximately 0.794%.

    Forsyth County, GA: The monthly growth rate is approximately 0.764%.

    b. Douglas County, CO: $P = 175,766(1.00854)^t$

    Loudoun County, VA: $P = 169,599(1.00794)^t$

    Forsyth County, GA: $P = 98,407(1.00764)^t$

    c. The doubling time for Douglas County can be found by letting P equal twice the April 1, 2000 population. The equation becomes $351,532 = 175,766(1.00854)^t$. Dividing each side by 175,766 results in $2 = 1.00854^t$. Take the log of each side of the equation to get: $\log(2) = \log(1.00854^t)$. Bring the exponent down in front, and divide through by $\log(1.00854)$ yields: $t = \dfrac{\log(2)}{\log(1.00854)} \approx 82$ months.

    Similarly, the doubling time for Loudoun County is $t = \dfrac{\log(2)}{\log(1.00794)} \approx 88$ months.

    Forsyth County has a doubling time of $t = \dfrac{\log(2)}{\log(1.00764)} \approx 91$ months.

19. Atmospheric CO_2 at Mauna Loa, Hawaii

   a. The annual growth rate is $r = 0.004$ or 0.4%.

   b. Solve the equation $400 = 312(1.004)^t$ to get $t \approx 62$ years. So about in the year 2021, we predict that the Mauna Loa CO_2 concentration will equal 400 ppm.

   c. For doubling time, solve $1.004^t = 2$ using logs. The result is $t \approx 174$ years.

   d. In general, the northern hemisphere experiences cooler weather from October to April, and warmer weather from April to October. From October to April, there is more plant decay than growth—this leads to a net increase in atmospheric CO_2. From April to October, the opposite happens—there is more growth than decay, and net atmospheric CO_2 concentration drops.

21. Galapagos Islands cactus finches

   a. See next graph. The data look somewhat exponential, although there is not a smooth decrease in population values.

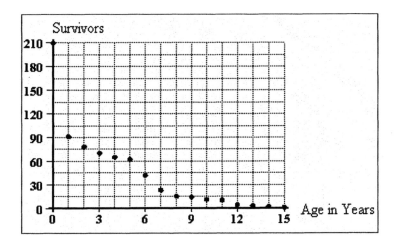

b.

Age	Log(survivors)	Age	Log(survivors)
0	2.32	8	1.18
1	1.96	9	1.15
2	1.89	10	1.04
3	1.85	11	1.00
4	1.81	12	0.60
5	1.79	13	0.48
6	1.62	14	0.30
7	1.36	15	0.00

c. See next graph. A line was drawn through the data; the points (0, 2.2) and (15, 0.4) were chosen on the line. The equation of the line through these points is $\log(y) = -0.12x + 2.2$.

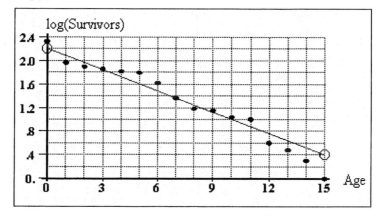

d. We use the formulas $M = 10^m$ and $y_0 = 10^b$ to find the exponential equation through the original data points. $M = 10^{-0.12} \approx 0.759$ and $y_0 = 10^{2.2} = 158$. The exponential equation is $y = 158(0.759)^x$.

e. The calculator's exponential regression equation is $y = 205(0.727)^x$. This equation has a significantly higher value for y_0 than does the straightedge method. The multiplier is similar, but slightly smaller than the one found in the straightedge method.

f. The annual death rate is $r = -0.273$ or -27.3%.

31

23. DDT in the food chain

a.

x	Y	y	$\log(y)$
1	0.04/1,000,000	4×10^{-8}	-7.4
2	0.23/1,000,000	2.3×10^{-7}	-6.6
3	2.07/1,000,000	2.07×10^{-6}	-5.7
4	13.8/1,000,000	1.38×10^{-5}	-4.9

b. The plot of the $(x, \log(y))$ points looks very linear. See next graph. An exponential model would be a very good fit to the original (x, y) data, because a linear model is a very good fit to the transformed data.

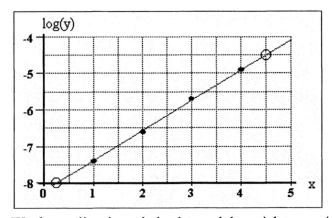

c. We draw a line through the data and then pick two points on the line (see previous graph). The points that we estimate on the line are (0.25, -8) and (4.5, -4.5). The line through these points has the equation $\log(y) = 0.8235x - 8.2059$. The

exponential model through the original data will be of the form $y = y_0 M^x$. The

value of M is $M = 10^{0.8235} \approx 6.66$, while the value of y_0 is

$y_0 = 10^{-8.2059} \approx 6.2 \times 10^{-9}$. The exponential equation is $y = 6.2 \times 10^{-9}(6.66)^x$,

where y is the DDT concentration and x the consumer level. To write the equation

with DDT concentrations expressed in units of parts per million, rewrite 6.2×10^{-9}

as 0.0062×10^{-6} or 0.0062 ppm. Then the equation becomes $y = 0.0062(6.66)^x$.

d. For each one-step increase in consumer level, DDT concentration increases by 6.66 or approximately a factor of 7.

e. The regression equation is $y = 0.0052(7.19)^x$, where y represents the DDT concentration in parts per million.

f. The correlation coefficient is $r = 0.999$, indicating that food-chain level is an excellent indicator of DDT concentration.

g. When $x = 5$ the DDT concentration would equal $y = 0.0052(7.19)^5 \approx 100$ ppm.

Chapter 6

1. Power Functions

   a.

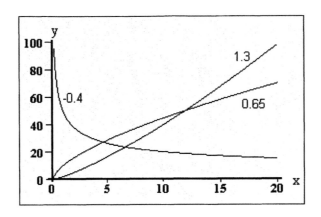

   b. The first two equations pass through the origin, but not the function with the negative power.

   c. A power function with a power c = 1 will result in a straight line.

3. Solving Power Functions

   a.

$$33 = 194x^{2.73}$$

$$\frac{33}{194} = x^{2.73}$$

$$\left(\frac{33}{194}\right)^{1/2.73} = \left(x^{2.73}\right)^{1/2.73}$$

$$0.523 = x$$

   b.

$$33 = 14.01x^{0.34}$$

$$\frac{33}{14.01} = x^{0.34}$$

$$\left(\frac{33}{14.01}\right)^{1/0.34} = \left(x^{0.34}\right)^{1/0.34}$$

$$12.426 = x$$

c.

$$33 = 23x^{-0.11}$$

$$\frac{33}{23} = x^{-0.11}$$

$$\left(\frac{33}{23}\right)^{\frac{1}{-0.11}} = \left(x^{-0.11}\right)^{\frac{1}{-0.11}}$$

$$0.038 = x$$

5. Phosphorus Discharge
   a. The phosphorus output for a watershed with an area of 25 mi² is 6,231 pounds per year. For a watershed 4 times as large in area, the phosphorus production is 18,629 pounds per year.
   b. *Answers may vary.* Using the graph, a phosphorus discharge of approximately 10,000 pounds per year comes from a drainage area of about 45.5 square miles.

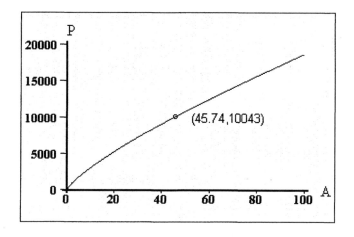

The area can also be calculated algebraically by solving for the unknown A:

$$10,000 = 490(A)^{0.79}$$

$$\left(\frac{10,000}{490}\right)^{\frac{1}{0.79}} = A$$

$$45.5\,\text{mi}^2 = A$$

7. Paleoclimatology
   a. A leaf with an area of 95 cm² lived in a climate with a mean annual precipitation of:

35

$$MAP = 26.753(95)^{0.547} = 323\,\text{cm}$$

This value is equivalent to 127 inches. *Answers may vary*: There are a few places in the U.S. where the mean annual precipitation exceeds 127 inches, including southeast Alaska, windward Hawai'i, and the Olympic Peninsula of Washington.

b. The average area of these two leaves is 20.5 square centimeters, corresponding to an expected mean annual precipitation of 140 centimeters:

$$MAP = 26.753(20.5)^{0.547} = 140\,\text{cm}$$

9. Glacier Dimensions

   a. Take the logarithms of the lengths and areas, and construct a scatterplot of the transformed data (see the following graph). Draw a line that best fits the data.

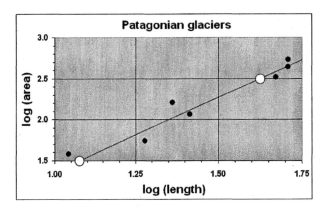

b. *Answers may vary.* Two data points (shown as open circles) lying on the line have coordinates (1.075, 1.5) and (1.625, 2.5). The slope is:

$$m = \frac{2.5 - 1.5}{1.625 - 1.075} = 1.818$$

The y-intercept can be calculated by substituting the slope and one of the pairs of values into the formula $\log(y) = b + m\log(x)$. We get:

$$1.5 = b + (1.818)(1.075)$$

$$-0.454 = b$$

The linear equation through the transformed data is:

$$\log(area) = -0.454 + 1.818 \log(length)$$

c. Because $c = m$ and $k = 10^b$, the corresponding power function is:

$$area = 0.35 \, length^{1.818}$$

d. The power of 1.818 is close to that of 2, matching the expectation of the glaciologists.

11. Australian Marsupials

a. Take the logarithms of the values for mass and time, and construct a scatterplot of the transformed data (see the following graph).

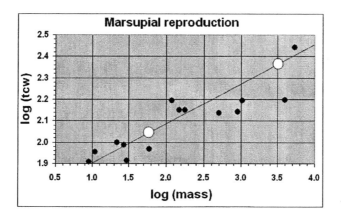

b. *Answers may vary.* Two points (shown as open circles) lying on the line have coordinates (1.8, 2.05) and (3.5, 2.37). The slope m is:

$$m = \frac{2.37 - 2.05}{3.5 - 1.8} = 0.188$$

The y-intercept can be calculated by substituting the slope and one of the pairs of value into the equation $\log(y) = b + m\log(x)$:

$$2.05 = b + (0.188)(1.8)$$
$$1.71 = b$$

The linear equation through the transformed data is:

$$\log(tcw) = 1.71 + 0.188\,\log(mass)$$

The corresponding power function is:

$$tcw = 51\,mass^{0.188}$$

c. The general shape of the power function is shown below. As the size of female dasyurids increases, so does the time from conception to weaning. For small dasyurids, a small change in mass yields a large change in time to weaning. But many large dasyurids have similar weaning times.

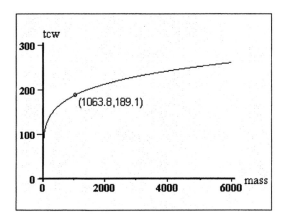

d. The expected time from conception to weaning for a red kangaroo is:

$$tcw = 51(3,500)^{0.188} \approx 237\,days$$

38

13. Makaopuhi Lava Lake
    a. The crusty layer floating on the lava lake thickens as a function of time, so time is the independent variable. The best fit regression model is:

$$T = 1.15762 \, (t)^{0.51699}$$

with time t in days and thickness T in feet. The correlation coefficient r is 0.9976. A power function is an excellent model of the data; time is an excellent predictor of the thickness of the crusty layer.
    b. When time $t = 1$, the model predicts a thickness $T = 1.2$ feet for the crust, which is a reasonable value, as we expect very thin crust after only one day of cooling.
    c. The model predicts a thickness of 47.8 feet for the crusty layer after 1,333 days, as opposed to the observed thickness of 52 feet. This prediction is still quite reasonable, as the actual data on which the model is based extend only to about 500 days. Extrapolating 800 days beyond the data and predicting a value that differs by only 10% from the actual measurement is a remarkable achievement.
    d. The geologists expect a "square root" behavior (a power of 0.5) and the model (power of 0.517) is consistent with that expectation. Cooling by conduction seems to be the dominant mechanism of heat loss.

15. Smallmouth Bass
    a. The best-fit power law regression between *age* (years) and *weight* (grams) is:

$$weight = 45.61 \, (age)^{1.555}$$

The correlation coefficient $r = 0.9036$.
    b. The best-fit regression is a fair model for the age and weight of smallmouth bass. The correlation coefficient is not that close to 1. There appears to be a large amount of scatter in the data; bass of the same age have several different weights. A linear or exponential model may also yield a reasonable correlation coefficient.
    c. The expected weight of a 2-year old bass is 134 grams according to the model:

$$134 = 45.61 \, (2)^{1.555}$$

17. Tonawanda Creek Discharge

   a. The minimum discharge was 100 cubic feet per second, and the maximum discharge was 939 cfs.

   b. *Results may vary.* 9 bins with a bin width of 100 cfs will enclose all the data. The sum of the frequencies yields a check on the calculation of the RCF for the first bin midpoint.

bin	frequency	BMP	RCF
100-200	17	150	31
200-300	6	250	14
300-400	2	350	8
400-500	2	450	6
500-600	1	550	4
600-700	2	650	3
700-800	0	750	1
800-900	0	850	1
900-1,000	1	950	1
Total	**31**		

   c. The data show a regular decrease in reverse cumulative frequency as discharge increases (as expected). The shape is consistent with a power function with a negative power.

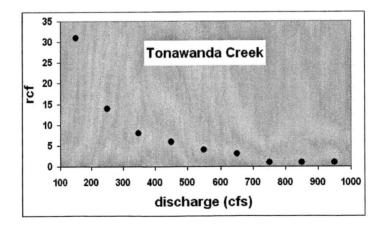

   d. The power law regression equation between reverse cumulative frequency *RCF* and the bin midpoint of discharge *BMP* is $RCF = 767,996\,BMP^{-1.9733}$. The power c for Tonawanda Creek is approximately -2, which falls in the interval -1 to -5.

e. The correlation coefficient r is -0.9738, with a negative sign that is consistent with a decreasing power function

f. Tonawanda Creek discharges for May, 2003 do show a power law frequency distribution. The correlation coefficient is very close to -1, indicating a very good fit to the data.

19. Landslides

a. *Results may vary!* 10 bins with a bin width of 2,000 m^2 embraced all the data.

BMP	RCF	BMP	RCF
1,000	3,163	11,000	259
3,000	1,909	13,000	196
5,000	1,300	15,000	141
7,000	1,028	17,000	80
9,000	652	19,000	35

b. The power law regression equation is $RCF = 160,191,465 \, BMP^{-1.4433}$. The correlation coefficient is -0.8976.

c. Landslides show a drop-off in frequency with increasing landslide area, however a power law is less than ideal as a model for this drop-off. The correlation coefficient is not close to -1.

d. The exponential regression equation that best fits the landslide data is $RCF = 4,485(0.9998)^{BMP}$. The correlation coefficient for the exponential model is -0.9929.

e. The exponential decay model seems to fit the data better, with a more favorable correlation coefficient.

21. River bank Failure

a.

BMP	RCF	log (BMP)	log (RCF)
2,500	187	3.40	2.27
7,500	75	3.88	1.88
12,500	43	4.10	1.63
17,500	21	4.24	1.32
22,500	16	4.35	1.20
27,500	9	4.44	0.95
32,500	5	4.51	0.70

b.

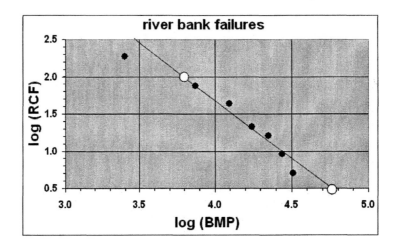

Two points on a line that represents the data have coordinates of (3.8, 2.0) and (4.75, 0.5). The slope of the line is:

$$m = \frac{0.5 - 2.0}{4.75 - 3.8} = -1.58$$

The y-intercept can be calculated by substituting the slope and one of the pairs of values into the equation $\log(y) = b + m\log(x)$:

$$2.0 = b + (-1.58)(3.8)$$
$$8.00 = b$$

The linear equation for the transformed data is:

42

$$\log(RCF) = 8.00 - 1.58 \log(BMP)$$

The corresponding power function is:

$$RCF = 10^8 \, BMP^{-1.58}$$

c. The corresponding reverse cumulative frequency for a bin interval of 40,000 m³ to 45,000 m³ (a bin midpoint of 42,500 m³) is approximately 4.9. The model suggests that about 5 river bank failures greater than equal to 40,000 m³ to 45,000 m³ will occur in a 37-year period. In other words, we would expect about 1 bank failure in this size range or greater approximately every 7 years.

Chapter 7

1. Finding sequence values
   a. 6   b. 54   c. –3   d. 36

3. Evaluating sequences A and B
   a. 42   b. –6   c. 0   d. 162

5. Sequence F

n	0	1	2	3	4	5
$f(n)$	10	18	26	34	42	50

Verbal: The initial term value is 10. To get any term value, add 8 to the previous term value. Math: $f(n) = f(n-1) + 8$ and $f(0) = 10$.

7. Sequence H

The initial term value is 2,000. To get any term value, multiply the previous term value by 0.75 then add 50.

n	0	1	2	3	4
$h(n)$	2,000	1,550	1,212.5	959.375	769.53125

9. Affine difference equation

   a.

n	0	1	2	3	4	5
$a(n)$	60	56	53.6	52.16	51.296	50.7776

   b.

n	0	1	2	3	4	5
$a(n)$	40	44	46.4	47.84	48.704	49.2224

   c.

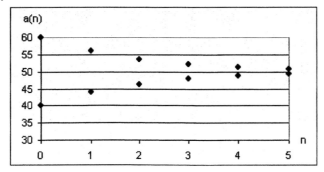

d. As *n* gets bigger, both sequences approach the number 50.

11. Finding solution equations
    a. Linear; solution equation is $u(n) = 50 + 10n$.

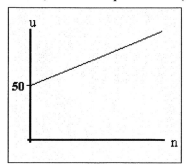

    b. Exponential; solution equation is $u(n) = 2,000(1.25)^n$

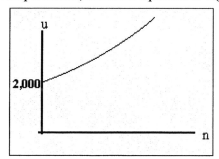

    c. Exponential; solution equation is $u(n) = 100(0.8)^n$

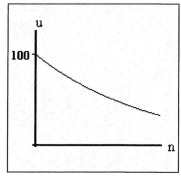

    d. Linear; solution equation is $u(n) = 25 - 3n$

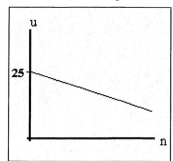

13. Consider $a(n) = a(n-1) + 5$

    a. If $a(0) = 10$, the solution equation is $a(n) = 10 + 5n$.

    b. If $a(0) = 15$, the solution equation is $a(n) = 15 + 5n$.

    c.

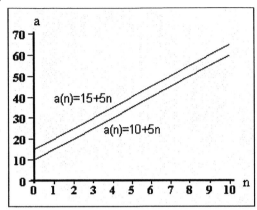

    d. Use the difference equation $a(n) = a(n-1) + 5$ and initial condition $a(0) = 10$ to generate all sequence values from $a(1) = 15$ to $a(4) = 30$. Alternately, use the solution equation and plug in $n = 4$ directly: $a(4) = 10 + 5(4) = 30$.

15. Modeling with affine difference equations

    a. Let $p(n)$ represent the population after n months. The affine difference equation is $p(n) = 1.05p(n-1) - 20$.

    b. After six months the population is approximately 2,544 individuals.

17. Meat consumption in China

This statistic tells us how meat consumption in any given year is related to meat consumption in the previous year. With this information we can write a difference equation: $c(n) = c(n-1) + 2$. In this equation, $c(n)$ represents meat consumption (in kg per person) in year n.

19. North Cascade glaciers

    a. $T(n) = T(n-1) - 0.30$

b.

Year	n	$T(n)$
1984	0	60
1985	1	59.7
1986	2	59.4
1987	3	59.1
1988	4	58.8
1989	5	58.5

c. Solution equation: $T(n) = 60 - 0.30n$. In 2001, when $n = 17$, the thickness is $T(17) = 54.9$ meters. The loss is 5.1 meters, which is 8.5% of its original 60 meters of thickness. If the length and width of the glacier have remained about the same during the 17-year period, then the volume would also decrease by about 8.5%. So the claim "about 10%" is reasonable.

21. Renewable energy

a. Industrialized: the yearly multiplier is $M = (1.27)^{1/10} \approx 1.02419$, and the yearly growth rate is $r = 0.02419 \approx 2.4\%$. Less industrialized: the yearly multiplier is $M = (1.19)^{1/10} \approx 1.01755$, and the yearly growth rate is $r = 0.01755 \approx 1.8\%$.

b. Industrialized: $E(n) = 1.02419\,E(n-1)$, $E(0) = 335,929$

Less industrialized: $E(n) = 1.01755\,E(n-1)$, $E(0) = 924,052$

c. Industrialized: $E(7) = 397,110$. Less industrialized: $E(7) = 1,043,727$

d. In about 155 years. This solution can be estimated from tables of values or graphs, or found algebraically if working with the solution equations. Factors that might make this assumption unlikely: politic changes, technological advances, etc.

23. Mexico City population

a. $p(n) = \big(p(n-1) + 0.75\big)(1.017)$

n	$p(n)$	n	$p(n)$
0	20	4	24.52
1	21.10	5	25.70
2	22.22	6	26.90
3	23.36		

b. The models are similar in that they both take into account the linear and exponential growth of the population, and they both do this as a two-stage process.

They are different in that they switch the order of these stages, which is simply an "accounting" difference. Neither modeling process is more correct (for human populations), both are making assumptions. The population actually changes all during the year, not at the end or the beginning of any given year.

25. Deer population

a. $P(n) = 1.04P(n-1) - 200$. This is an affine difference equation.

b.

n	$P(n)$	n	$P(n)$	n	$P(n)$	N	$P(n)$
0	2,000	4	1,490	8	894	12	197
1	1,880	5	1,350	9	730	13	5
2	1,755	6	1,204	10	559	14	-195
3	1,625	7	1,052	11	382		

c.

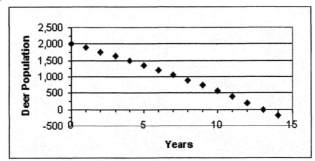

d. 80 deer each season. Then the 4% growth (4% of 2,000 = 80) will be balanced out by the 80 killed.

e. 46 to 47 deer

Chapter 8

1. Solve the system $60 = a + b$ and $58.634 = 1.025a + b$

   Rewrite the first equation as $b = 60 - a$, and substitute into the second equation to get $58.634 = 1.025a + 60 - a$. Collect like terms to produce $-1.366 = 0.025a$. Divide each side by 0.025 to yield the solution for a: $a = -54.64$. Substitute to find the value of b: $b = 60 - (-54.64) = 114.64$.

3. Affine solution equations

   a. The solution equation is of the form: $p(n) = a(2.5)^n + b$. Substituting the point $(n, p) = (0, 400)$ yields the equation: $400 = a + b$. The point $(n, p) = (1, 1900)$ yields a second equation: $1,900 = 2.5a + b$.

   The first equation can be written as $b = 400 - a$. Substitute this into the second equation to get $1,900 = 2.5a + 400 - a$ which simplifies into: $1,500 = 1.5a$. The solution to this equation is $a = 1,000$. If $a = 1,000$, then $b = 400 - 1,000 = -600$. The solution equation is $p(n) = 1,000(2.5)^n - 600$.

   b. The solution equation is of the form: $p(n) = a(0.75)^n + b$. Substitute the point $(0, 510)$ to yield the equation: $510 = a + b$. Similarly, the point $(1, 337.5)$ yields the equation: $337.5 = 0.75a + b$.

   Rewrite the first equation as $b = 510 - a$, and substitute into the second equation to get $337.5 = 0.75a + 510 - a$ which simplifies to $-172.5 = -0.25a$. The solution to this equation is $a = 690$. For this value of a, $b = 510 - 690 = -180$. The solution equation is: $p(n) = 690(0.75)^n - 180$.

   c. Start with the two points $(0, 75)$ and $(1, 2.25)$. Substitute these points into $p(n) = a(1.03)^n + b$ to yield the two equations $a + b = 75$ and $1.03a + b = 2.25$. Solve these two equations to obtain $a = -2,425$ and $b = 2,500$; the solution equation is $p(n) = -2,425(1.03)^n + 2,500$.

5. Finding solution equations

   a. This is an exponential difference equation, with a multiplier of $M = 0.64$. The y-intercept is $y_0 = 4$. The general form of an exponential equation is $y(x) = y_0 M^x$, so the solution equation is $u(n) = 4(0.64)^n$.

   b. This is a linear difference equation with a slope of $m = -6$. The y-intercept is $b = 100$. The general form of a linear equation is $y(x) = b + mx$, so the solution equation is $v(n) = 100 - 6n$.

   c. This is an affine difference equation. The solution equation is of the form $w(n) = a(0.2)^n + b$. Substitute the two points $(0, 7)$ and $(1, 5.4)$ to get the equations $a + b = 7$ and $0.2a + b = 5.4$. Solving this system of two equations produces the solution equation $w(n) = 2(0.2)^n + 5$.

7. Population model

   a. The initial condition $p(0) = 600$ corresponds to the point $(0, 600)$. Use the difference equation to obtain the second point $(1, 680)$. Substitute these points into $p(n) = a(0.80)^n + b$ to yield $a + b = 600$ and $0.80a + b = 680$. Solve this system of equations to get $a = -400$ and $b = 1,000$. The solution equation is

   $p(n) = -400(0.80)^n + 1,000$.

   The initial condition $p(0) = 800$ corresponds to the point $(0, 800)$. Use the difference equation to obtain the second point $(1, 840)$. Substitute these points into $p(n) = a(0.80)^n + b$ to yield $a + b = 800$ and $0.80a + b = 840$. Solve this system of equations to get the solution equation $p(n) = -200(0.80)^n + 1,000$.

   Use the same approach described previously with $p(0) = 1,300$ to get the solution equation $p(n) = 300(0.80)^n + 1,000$.

b.

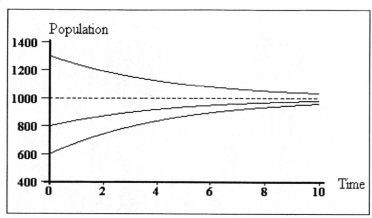

c. The equilibrium value is stable, because all populations move toward that value.

9. Finding equilibrium values

a. Solve $E = 2.2E - 48$ to get $E = 40$. Check: $40 = 2.2(40) - 48$ ✓

b. Solve $E = 3E$. The solution is $E = 0$. Check: $0 = 3(0)$ ✓

c. Solve $E = 0.75E + 20$ to obtain the equilibrium value $E = 80$. Check:
   $80 = 0.75(80) + 20$ ✓

d. Solve $E = E + E^2 - 25$. First simplify to get $0 = E^2 - 25$, and then factor to obtain
   $0 = (E - 5)(E + 5)$. Set each factor equal to zero to yield the two equilibrium values
   $E = \pm 5$.

   Check: $5 = 5 + 5^2 - 25$ ✓ Check: $-5 = -5 + (-5)^2 - 25$ ✓

11. Classifying equilibrium values

a. Solve $E = 1.25E - 2{,}000$ to yield $E = 8{,}000$. Use initial conditions slightly larger
   than or smaller than 8,000. Sequence values move *away from* the equilibrium
   value of 8,000, thus the equilibrium value is unstable.

b. Solve $E = 0.5E + 60$ to get $E = 120$. Use initial conditions slightly larger than or
   smaller than 120. Sequence values move *toward* the equilibrium value of 120, thus
   the equilibrium value is stable.

13. Equilibrium values and affine solution equations

a. Solve $E = 1.5E - 45$. The equilibrium value is $E = 90$, which is also the value for
   b. Because $a + b = 300$, we can solve to find that $a = 210$. The solution equation is
   $u(n) = 210(1.5)^n + 90$.

b. Solve $E = 0.98E + 40$. The equilibrium value is $E = 2,000$, which is also b. Because $a + b = 102$, we know that $a = -1,898$. The solution equation is

$$v(n) = -1,898(0.98)^n + 2,000.$$

c. The equilibrium value is $E = -8,000$. The solution equation is

$$w(n) = 10,000(1.025)^n - 8,000.$$

15. Forest plantation

    a. Let T equal the number of live trees, and n the number of years. The difference equation is $T(n) = 0.75T(n-1) + 500$.

    b. Use the points $(0, 6000)$ and $(1, 5000)$. The solution equation is of the form $T(n) = a(0.75)^n + b$. Substitute the two points to create the equations $a + b = 6,000$ and $0.75a + b = 5,000$. Solve these two equations to get $a = 4,000$ and $b = 2,000$. The solution equation is $T(n) = 4,000(0.75)^n + 2,000$.

    c.

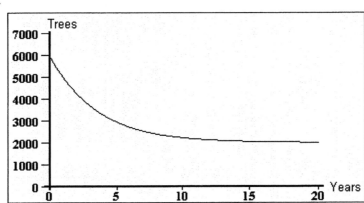

    d. The tree population will level off at 2,000 trees. When that level is reached, there will be (25%)(2,000) = 500 trees harvested each year.

17. Microloan example

    a. Solve $E = 1.0025E - 50$ to obtain the equilibrium value of $E = 20,000$. In terms of the loan, if the loan balance in one month is $20,000, then the loan balance will remain at $20,000 for the following month. This can be explained by noting that the monthly payment of $50 would only pay the interest on borrowing $20,000 at a 0.25% rate. In this example, only $1,500 was borrowed by the community—well below the $20,000 equilibrium value. With an initial loan amount of $1,500, the

$50 monthly payment will cover not only the monthly interest, but also some of the principal each month.

b. The solution equation is of the form $u(n) = a(1.0025)^n + b$. The value of b is the same as the equilibrium value, so $b = 20,000$. Because $a + b$ is equal to the initial condition, the value of a is $a = -18,500$. Thus the solution equation is

$u(n) = -18,500(1.0025)^n + 20,000$.

c. Using the difference equation, we create a table of values displaying the balance each month. After 10 months the balance is $1,032.26. See the following table.

n	$u(n)$	N	$u(n)$
0	1500	6	1220.76
1	1453.75	7	1173.81
2	1407.38	8	1126.75
3	1360.90	9	1079.56
4	1314.31	10	1032.26
5	1267.59		

Using the solution equation, we find that the balance after 10 months is $-18,500(1.0025)^{10} + 20,000 = 1,032.26$, which is what the difference equation predicts.

19. Amalgam fillings

a. The multiplier raised to the 15th power is equal to ½ or 0.5. Thus we can solve the equation $M^{15} = 0.5$ to get $M = (0.5)^{1/15} \approx 0.9548$.

b. Each year the body will absorb about $120 \times 365 = 43,800$ micrograms. This is equivalent to $43,800 \times 10^{-6}$ grams, or 0.0438 grams.

c. Let A represent the amount of mercury in the body in micrograms, and n the number of years. The difference equation is $A(n) = 0.9548 A(n-1) + 43,800$.

d. Solve $E = 0.9548E + 43,800$ to produce the equilibrium value of $E = 969,026.5487 \approx 969,000$ micrograms. Using initial conditions slightly smaller and slightly larger than the equilibrium value, all sequences converge toward the equilibrium value; thus the equilibrium value is stable. Stability implies that no matter what the current level of mercury in the body, the long-term mercury level will approach 969,000 micrograms.

e. The initial quantity of 0.5 grams is equal to 500,000 micrograms. The value of b is equal to the equilibrium value of 969,000. Because $a+b=500,000$, the value of a is $a=-469,000$. The solution equation is $A(n)=-469,000(0.9548)^n+969,000$.

21. Rework salmon hatcheries example

    The initial condition and difference equation describing the salmon population are $p(0)=2,000$ and $p(n)=0.75p(n-1)+k$. The equilibrium value is found by solving the equation: $E=0.75E+k$ which results in $E=4k$. The affine solution equation is $p(n)=(2,000-4k)(0.75)^n+4k$.

    Now use the fact that in 15 years the salmon population is required to total 6,000. We get: $6,000=(2,000-4k)(0.75)^{15}+4k$. This equation can be solved for k to yield $k=1,513.54$ salmon. The hatchery must add slightly more than 1,500 salmon each year.

23. Grameen Shakti microenergy

    a. $u(0)=\$382.50$

    b. $u(n)=1.01u(n-1)-k$

    c. Solve $E=1.01E-k$ to produce the equilibrium value of $E=100k$.

    d. The solution equation is $u(n)=(382.50-100k)(1.01)^n+100k$

    e. Solve the equation $0=(382.50-100k)(1.01)^{36}+100k$. Expand the equation to get $0=382.50(1.01)^{36}-100(1.01)^{36}k+100k$. Factor out k, and move all numbers to the left: $-382.50(1.01)^{36}=\left(100-100(1.01)^{36}\right)k$. Isolate k to get

    $$k=\frac{-382.50(1.01)^{36}}{100-100(1.01)^{36}}=12.7044735.$$ The monthly payment is $k=\$12.70$.

    f. The amount due after 35 months is approximately $12.77. The last (36th) payment will be equal to $12.77 + interest charges: $12.77 + 0.01(\$12.77)=\$12.89$.

Chapter 9

1. Graphing a logistic difference equation

   a. The carrying capacity is $K = 1,000$, the unrestricted growth rate is $r_{max} = 9\%$.

   b.

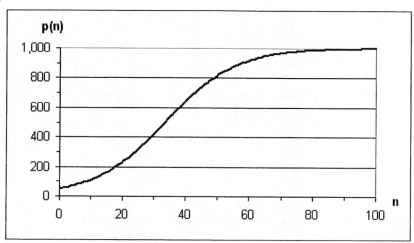

3. Writing logistic difference equations

   a. $p(n) = 1.02 p(n-1) - \dfrac{0.02}{60} p(n-1)^2$

   b. $p(n) = 1.075 p(n-1) - \dfrac{0.075}{4,000} p(n-1)^2$

   c. $p(n) = 1.05 p(n-1) - \dfrac{0.05}{20,000} p(n-1)^2$

   d. $p(n) = 1.003 p(n-1) - \dfrac{0.003}{2,000} p(n-1)^2$

5. Graph in Figure 9-3

   a. Slope $= -0.05/5,000,000 = -0.00000001 = -10^{-8}$

   b. The equation of the line is in the form $y = -10^{-8}x + b$ or $r = -10^{-8}p + b$.

   Substituting the first point $(p,r) = (10^6, \, 0.05)$ results in $0.05 = -10^{-8}(10^6) + b$. This simplifies to $0.05 = -0.01 + b$. Solving for b yields $b = 0.06$. Thus the equation of the line is $r = 0.06 - 0.00000001p$.

c. The y-intercept is 0.06 or 6%, which is the unrestricted growth rate (r_{max}). This agrees exactly with the graph in Figure 9-3.

d. Yes, if the fraction used for the slope is converted to a decimal, then the text's formula agrees with the one found in part b).

7. Linearly decreasing population growth rates

a.

$p =$ Population	$r =$ Growth Rate
0	0.1250
100	0.1125
200	0.1
300	0.0875
400	0.075
500	0.0625
600	0.05
700	0.0375
800	0.025
900	0.0125
1,000	0

b. The unrestricted growth rate is $r_{max} = 0.125$ or 12.5%. This is the growth rate when the population equals 0.

c. $K = 1,000$ is the carrying capacity. In logistic growth, the carrying capacity is the population when the growth rate equals 0.

d. $r = 0.125 - \dfrac{0.125}{1,000}p$

9. World population

a. The carrying capacity of earth is projected to be about 10 billion people. *Answers may vary.*

b. With $r_{max} = 0.016$ and $K = 10$, the logistic difference equation is

$$p(n) = 1.016p(n-1) - \frac{0.016}{10}p(n-1)^2 .$$

c.

Year	n	p	Year	n	p
1850	0	1.26	2030	180	7.19
1880	30	1.89	2060	210	8.05
1910	60	2.73	2090	240	8.70
1940	90	3.77	2120	270	9.16
1970	120	4.94	2150	300	9.46
2000	150	6.12			

d. Scrolling through a table of values, we find that when $n = 305$ (the year 2155) the world's population will be approximately $p = 9.5$ billion people.

11. G.F. Gause bacteria experiment

a.

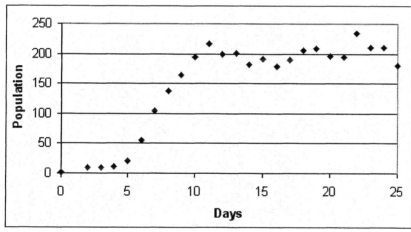

b. The equilibrium is about 200 bacteria.

c. The logistic difference equation is $p(n) = 1.8p(n-1) - \dfrac{0.8}{200}p(n-1)^2$.

d. Gauss was referring to the last 10-13 days in which the graph of the population bounces up and down between 180 and 230 bacteria. It's possible that the test tube environment was not held perfectly steady—too much or too little food or water, or the waste products were not removed uniformly each day. Any of these conditions might lead to a fluctuating population.

e. The population would outgrow the food supply, and the waste materials would pile up, so that the population would eventually start to die off. It's hard to know whether the population would have reached the equilibrium level before the die-off or not. If it did, the graph might look like the following.

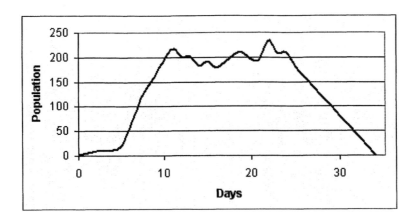

13. Population of England and Wales

a. After drawing a line of best fit, we find that when the population is 0, the annual growth rate is $r_{max} = 0.02$ (the vertical intercept). When the annual growth rate equals 0, the carrying capacity is $K = 60$ million people. *Answers may vary.*

b. The equation is $p(n) = 1.02p(n-1) - \dfrac{0.02}{60}p(n-1)^2$, where p represents the population in millions, and n the number of years.

c.

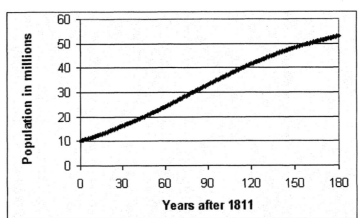

d. During the industrial revolution, many people from rural areas moved to the city. City families had fewer children than rural families because of the lesser need for labor to run family affairs (farms required lots of "hands"). There was also a great increase in the general level of education, and in the number of women that joined the working ranks. Both of these factors contribute to declining fertility rates.

15. Finding equilibrium values

a. Set up the equation: $E = 1.15E - \dfrac{0.15}{60}E^2$.

Subtract E from each side: $0 = 0.15E - \dfrac{0.15}{60}E^2$.

Factor out E: $0 = E\left(0.15 - \dfrac{0.15}{60}E\right)$

Set each factor equal to 0: $E = 0$ or $0.15 - \dfrac{0.15}{60}E = 0$

The second equation has the solution: $E = 60$

The two equilibrium values are $E = 0$ and $E = 60$.

b. Set up the equation: $E = 1.15E - \dfrac{0.15}{60}E^2 - 2$

Subtract E from each side: $0 = 0.15E - \dfrac{0.15}{60}E^2 - 2$

Multiply each side by $\dfrac{60}{0.15}$ to clear the denominator and any decimals:

$$\left(\dfrac{60}{0.15}\right)0 = \left(\dfrac{60}{0.15}\right)\left(0.15E - \dfrac{0.15}{60}E^2 - 2\right)$$

$$0 = 60E - E^2 - 800$$

Simplify: $E^2 - 60E + 800 = 0$

Factor: $(E - 20)(E - 40) = 0$

Set each factor equal to 0: $E = 20$ and $E = 40$

17. Bacteria population growing logistically

a.

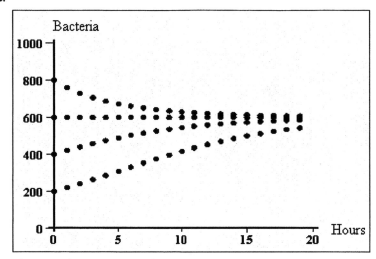

b. The carrying capacity of the test tube is 600 bacteria.

c. Solve: $E = 1.15E - 0.00025E^2$

$0 = 0.15E - 0.00025E^2$

$0 = E(0.15 - 0.00025E)$

$E = 0$ or $0.15 - 0.00025E = 0$

$E = 0$ or $E = 600$

If the bacteria population is at either equilibrium value, then the population will stay at that level. The second equilibrium value, $E = 600$, is the carrying capacity.

An alternative way to find the (non-zero) equilibrium level is to find the carrying capacity K directly from the difference equation. The equation indicates that

$r_{max} = 0.15$ and that $\dfrac{r_{max}}{K} = 0.00025$. Combine these two equations to get

$\dfrac{0.15}{K} = 0.00025$. Multiply each side by K to get $0.15 = 0.00025K$. Isolate K to

yield the carrying capacity (an equilibrium level) of $K = 600$ bacteria.

19. Fixed harvest of elk

a. Set up the equation: $E = 1.30E - \dfrac{0.30}{1,500}E^2 - 50$.

Move all terms to the left side of the equation to get $\dfrac{0.30}{1,500}E^2 - 0.30E + 50 = 0$.

60

Multiply each side of the equation by $\dfrac{1,500}{0.30}$ to clear the denominator and all

decimals. The result is $E^2 - 1,500E + 250,000 = 0$.

This equation will not factor, so use the quadratic formula with $A = 1$, $B = -1,500$,
and $C = 250,000$. The solutions are:

$$E = \frac{1,500 \pm \sqrt{(-1,500)^2 - 4(1)(250,000)}}{2(1)}$$

$$E = \frac{1,500 \pm \sqrt{1,250,000}}{2}$$

$E \approx 1,309$ or $E \approx 191$

b.

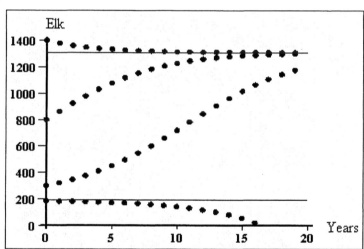

c. The upper equilibrium value, $E \approx 1,309$ elk, is a stable equilibrium value. The
lower equilibrium value, $E = 191$, is unstable. The lower equilibrium value may
tell us that if the elk population falls below 190, then the population will die off
(this is assuming that 50 elk are killed each year). This might be caused by
decreased reproduction options, or decreased genetic diversity.

21. Pacific halibut population fixed harvest

a. $u(n) = 1.71u(n-1) - \dfrac{0.71}{80}u(n-1)^2 - 3$

b. The population in the long term level offs at about 75.5 million kg of halibut.

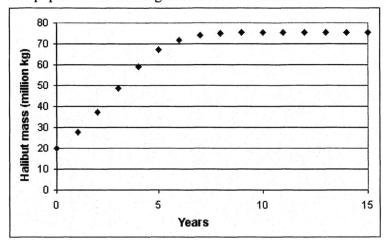

c. 75.5 kg is 4.5 kg below 80 kg. As a percentage, this is 4.5 kg/80 kg = 5.625% below the carrying capacity.

23. Pacific halibut population proportional harvest

a. The difference equation is $u(n) = 1.71u(n-1) - \dfrac{0.71}{80}u(n-1)^2 - 0.15u(n-1)$, which

can be simplified as $u(n) = 1.56u(n-1) - \dfrac{0.71}{80}u(n-1)^2$.

b. The population mass levels off at about 63.1 million kg .

c. Using trial and error, we find that when the proportional harvest is about 9%, then the graph levels off at 70 million kg.

d. A general logistic harvest equation for this exercise has the form

$u(n) = 1.71u(n-1) - \dfrac{0.71}{80}u(n-1)^2 - Hu(n-1)$, where H is the rate of harvest. In

this equation substitute $E = 70$ for $u(n)$ and $u(n-1)$. The result is

$70 = 1.71(70) - \dfrac{0.71}{80}(70)^2 - (H)(70)$. Solving for H yields the exact proportional

harvest: $H = 8.875\%$.

25. Periodic behavior of an insect population

a. The difference equation is $p(n) = 3.3p(n-1) - \dfrac{2.3}{1,000}p(n-1)^2$.

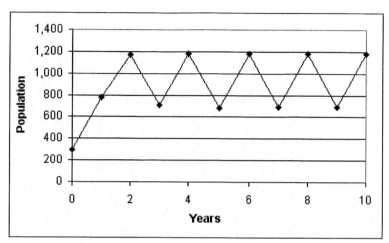

b. The population cycles between the approximate values of 688 and 1,182 insects.

27. Sensitivity to initial conditions

a. The difference equation is $p(n) = 4.0p(n-1) - \dfrac{3.0}{100}p(n-1)^2$, where p is the population of insects, and n is the number of years.

b.

n	Population	Population	Population
0	79	80	81
1	128.8	128.0	127.2
2	17.6	20.5	23.5
3	61.2	69.3	77.5
4	132.4	133.1	129.8
5	3.6	0.9	13.6
6	13.9	3.4	49.0
7	49.8	13.2	123.9
8	124.8	47.7	35.0
9	31.8	122.5	103.2
10	96.9	39.7	93.2

c. The model is sensitive to initial conditions. Small differences in initial population values create large differences in the projected population values, even by $n = 6$.

Chapter 10

1. Regarding Example 10-1

$a(3) = 4b(2) + 3c(2)$	$b(3) = 0.3a(2)$	$c(3) = 0.1b(2)$
$= 4(210) + 3(3)$	$= 0.3(150)$	$= 0.1(210)$
$= 849$	$= 45$	$= 21$
$a(4) = 4b(3) + 3c(3)$	$b(4) = 0.3a(3)$	$c(4) = 0.1b(3)$
$= 4(45) + 3(21)$	$= 0.3(849)$	$= 0.1(45)$
$= 243$	$= 254.7$	$= 4.5$
$a(5) = 4b(4) + 3c(4)$	$b(5) = 0.3a(4)$	$c(5) = 0.1b(4)$
$= 4(254.7) + 3(4.5)$	$= 0.3(243)$	$= 0.1(254.7)$
$= 1,032.3$	$= 72.9$	$= 25.47$

n	$a(n)$	$b(n)$	$c(n)$	Total
0	100	100	100	300
1	700	30	10	740
2	150	210	3	363
3	849	45	21	915
4	243	254.7	4.5	502.2
5	1,032.3	72.9	25.47	1,130.67

3. System of three difference equations

   Hint: Find all three populations in any given year before moving on to the next year.

n	$a(n)$	$b(n)$	$c(n)$
0	20	10	30
1	$20 + 10 = 30$	$10 + 30 = 40$	$20 + 30 = 50$
2	$30 + 40 = 70$	$40 + 50 = 90$	$30 + 50 = 80$
3	$70 + 90 = 160$	$90 + 80 = 170$	$70 + 80 = 150$

5. System of four difference equations

Hint: Find all four populations in any given year before moving on to the next year.

n	$a(n)$	$b(n)$	$c(n)$	$d(n)$
0	2	5	3	4
1	$2(3)+1=7$	$3(4)+1=13$	$2(2)-2=2$	$3(5)-2=13$
2	$2(2)+1=5$	$3(13)+1=40$	$2(7)-2=12$	$3(13)-2=37$
3	$2(12)+1=25$	$3(37)+1=112$	$2(5)-2=8$	$3(40)-2=118$

7. Population with three age groups

   a. Every year, each member in group A will produce 2 offspring that survive for one full year. Group B members will each produce 3 offspring every year that survive for a full year. Group C members do not produce any offspring that survive a full year.

   b. Each year, 60% of group A and 25% of group B will live one additional year. No members in group C will live one additional year.

   c.

	Group A	Group B	Group C
Annual Fertility Rate	2	3	0
Annual Survival Rate	0.6	0.25	0

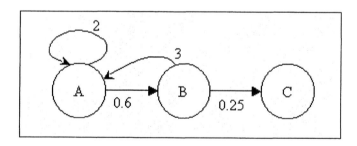

   d. *Hint*: The fertility and survival rates are annual rates. To write the equations, it may be helpful to picture the arrows in the life-cycle graph as starting in year $n-1$ and finishing in year n. Thus the equations are:

   $a(n) = 2a(n-1) + 3b(n-1)$
   $b(n) = 0.6a(n-1)$
   $c(n) = 0.25b(n-1)$

9. Life cycle and system of difference equations

Hint: To write the equations, it may be helpful to picture the arrows in the life-cycle graph as starting in year $n-1$ and finishing in year n. Thus the equations are:

$$a(n) = 2.0b(n-1) + 3.5c(n-1)$$
$$b(n) = 0.36a(n-1)$$
$$c(n) = 0.25b(n-1) + 0.88c(n-1)$$

11. Sand smelt

a. Each female in age group A produces 1,423 eggs. Half of these are female, so there are 711.5 female eggs. Only 0.16% of these survive to be counted in age group A at the end of the first year: $0.16\% \times 711.5 = 1.1384$. Thus the fertility rate for females in group A is about 1.14. Similar computations result in a 3.65 fertility rate for group B, and a 6.97 fertility rate for group C.

b.

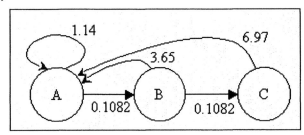

$$a(n) = 1.14a(n-1) + 3.65b(n-1) + 6.97c(n-1)$$
$$b(n) = 0.1082a(n-1)$$
$$c(n) = 0.1082b(n-1)$$

c. Population values for each age group are derived using technology. Sum across each row to get the total female smelt in any given year. *Total column values may vary depending on the rounding strategy employed.*

n	$a(n)$	$b(n)$	$c(n)$	Total
0	1,000	1,000	1,000	3,000
1	11,760	108	108	11,976
2	14,555	1,272	12	15,840
3	21,319	1,575	138	23,032
4	31,012	2,307	170	33,489
5	44,961	3,355	250	48,566
6	65,243	4,865	363	70,470
7	94,664	7,059	526	102,249

d. $M = 70,470/48,566 = 1.451$, which makes $r = 45.1\%$

e. $M = 102,249/70,470 = 1.451$, which makes $r = 45.1\%$. It appears that the sand smelt population's growth rate is stabilizing at 45.1%. This should be the long-term growth rate.

f. Using the values for $n = 7$, the percentages in groups A, B, and C in the long term are $94,664/102,249 = 92.6\%$, $7,059/102,249 = 6.9\%$ and $526/102,249 = 0.5\%$ respectively.

g. The revised fertility rate for group A is $(0.60)(1.14) = 0.68$. Similarly, for groups B and C the rates are 2.19 and 4.18, respectively. To determine the stable age distribution, we inspect population values in, for example, year $n = 7$. See the following table.

n	A	B	C	Total
7	4,939	548	61	5,548

The percentages in groups A, B, and C in the long term are $4,939/5,548 = 89.0\%$, $548/5,548 = 9.9\%$ and $61/5,548 = 1.1\%$ respectively. *Answers may vary.*

13. PCH revisited

Year	Calves	Yearlings	Adults
$n = 20$	153.500	126.579	422.236
$n = 21$	173.117	142.755	476.194
Growth Rate	$\dfrac{173.117 - 153.500}{153.500} = 12.8\%$	$\dfrac{142.755 - 126.579}{126.579} = 12.8\%$	$\dfrac{476.194 - 422.236}{422.236} = 12.8\%$

Year	Calves	Yearlings	Adults
$n = 30$	510.977	421.362	1,405.552
$n = 31$	576.276	475.209	1,585.171
Growth Rate	$\dfrac{576.276 - 510.977}{510.977} = 12.8\%$	$\dfrac{475.209 - 421.362}{421.362} = 12.8\%$	$\dfrac{1,585.171 - 1,405.552}{1,405.552} = 12.8\%$

All the growth rates equal 12.8%, which means that the long-term growth rates for each stage will stabilize at this value. We also know that the total population will increase, in the long term, by 12.8% annually.

15. Waterbuck population

a. $a(n) = 0.048b(n-1) + 0.081c(n-1)$

$b(n) = 0.94a(n-1)$

$c(n) = 1.00b(n-1) + 0.75c(n-1)$

b. A male:female ratio of 1:2.24, means that out of every 3.24 waterbuck, 2.24 are female. Thus the percentage of females is $2.24/3.24 = 69.1\%$. Approximately $(69.1\%)(494) = 342$ of the waterbuck sighted by air were female.

c. The initial conditions are $a(0) = 114$, $b(0) = 114$, $c(0) = 114$.

d. See population values in the following table for years $n = 9$ and $n = 10$. *Answers may vary depending on years chosen.*

Year	Calves	Yearlings	Adults
$n = 9$	8.29	9.07	82.63
$n = 10$	7.13	7.79	71.04
Decay Rate	$\frac{7.13-8.29}{8.29} \times 100\% \approx -14\%$	$\frac{7.79-9.07}{9.07} \times 100\% \approx -14\%$	$\frac{71.04-82.63}{82.63} \times 100\% \approx -14\%$

In the long term, each growth stage in the population decreases by about 14.0% annually. Thus the total population will do the same. The model predicts that after about $n = 25$ (year 2000) the total population would number under 10 waterbuck. When a herd gets this small, the chances for recovery are greatly diminished. We could assume that the waterbuck would go extinct in the park.

17. Pollution in Lake Erie and Lake Ontario

a. In year $n = 2$, Lake Erie had a pollution level of 0.4489. Lake Erie will flush 33% of this amount, or $0.33(0.4489) = 0.148137$.

b. In year $n = 2$, Lake Ontario's pollution level was 1.1839. Lake Ontario will lose 17% of this amount, or $0.17(1.1839) = 0.201263$.

c. In year $n = 3$, Lake Erie's pollution level will equal $0.4489 - 0.148137 = 0.300763$.

d. In year $n = 3$, Lake Ontario's pollution level will equal $1.1839 - 0.201263 + 0.148137 = 1.130774$.

e. These answers match the results in the table.

n	Erie	Ontario
0	1	1
1	0.67	1.16
2	0.4489	1.1839
3	0.300763	1.130774

19. Human exposure to lead

a.

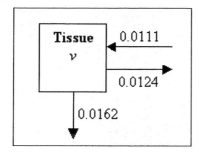

Tissue: $v(n) = 0.9714v(n-1) + 0.0111u(n-1)$

b.

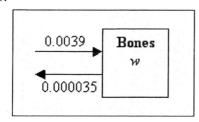

Bones: $w(n) = 0.999965w(n-1) + 0.0039u(n-1)$

c. The initial conditions are $u(0) = 0\ \mu g$, $v(0) = 0\ \mu g$ and $w(0) = 0\ \mu g$.

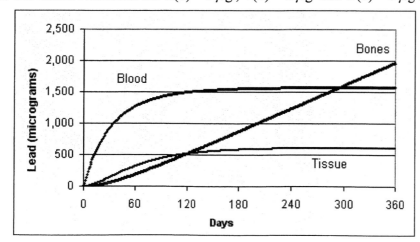

d. The amounts of lead in the blood and the tissue grow quickly, but then gradually level off. It appears that each of these amounts will reach their own equilibrium level. The amount of lead in the bones grows slowly at first, and then climbs in what appears to be a linear manner. Eventually the amount of lead in the bones exceeds that of the blood and tissue; this is due to the fact that the bones only lose lead to the blood, and the flow rate is extremely slow (0.0035% each day).

e. After 360 days, the blood lead level concentration was:

$$\frac{1,577\,\mu g}{5\,L} \cdot \frac{1\,L}{10\,dL} = 31.54\,\frac{\mu g}{dL}.$$ Yes, the blood lead level was dangerously high.

21. Measles in Nigeria

a. $C(n) = 0.00003\,S(n-1)\,C(n-1)$

b. $S(n) = S(n-1) - 0.00003\,S(n-1)\,C(n-1) + 360$

c. $C(0) = 20$; $S(0) = 30,000$

d.

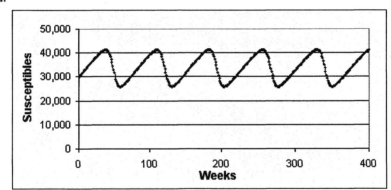

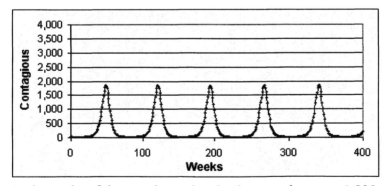

e. At the peaks of the measles outbreaks there are between 1,830 and 1,840 contagious children. The peaks occur when $n = 48, 121, 195, 268,$ and 341 weeks; thus the frequency of measles outbreaks is 73 to 74 weeks.

70

Chapter 11

Note: If using technology other than the TI-83/84 to compute descriptive statistics, your results may differ slightly for Q1, Q3, boxplots and the Bowley skew.

1. Discarded tires

    a. The mean is the sum of the 5 sample measurements, divided by 5. The median is the 3rd measurement when the data are arranged in ascending (or descending) order.

	Fe	Mn	Al	Cr	Cu	Zn
Mean (mg/L)	6.984	0.732	0.336	0.094	0.182	1.012
Median (mg/L)	7.10	0.71	0.36	0.07	0.16	1.06

    b. The range is the maximum minus the minimum.

	Mn	Cu	Cr	Al	Zn	Fe
Range (mg/L)	0.08	0.14	0.17	0.33	0.43	0.96

    The ranges indicate the amount of variation in each data set. The manganese data have the least amount of variation, whereas the iron data have the most.

    c. 60% (3 out of 5) of the samples have chromium concentrations below 0.10 mg/L.

3. Land biomass

    Multiply the area percentage by the biomass to obtain the weighted biomass for each land type. Sum the weighted biomass values to get the mean biomass for Earth: approximately 3.8 kg C/m^2.

Land Type	Area Percentage	Biomass (kg C/m^2)	Weighted Biomass (kg C/m^2)
Forest	29%	10.0	2.900
Grassland	27%	2.3	0.621
Desert	13%	0.3	0.039
Tundra	8%	0.8	0.064
Wetland	2%	2.7	0.054
Cultivated	11%	1.4	0.154
Rock and Ice	10%	0.0	0.000
		Mean Biomass:	$\Sigma = 3.832$

5. U.S. poverty rates

a. Given: $\text{poverty rate} = \dfrac{\#\text{ in poverty}}{\#\text{ in race}}$. Rearrange to get: $\#\text{ in race} = \dfrac{\#\text{ in poverty}}{\text{poverty rate}}$.

Use this last formula to find the number of people in each race, and then sum the results. The total of 283 million people is about correct for the U.S. in years 2000-2001. See the following table.

Race	Number in Race	% of U.S. Population
White (Non-Hispanic)	194,986,842	68.89%
Black	35,659,292	12.60%
American Indian and Alaska Native	3,226,667	1.14%
Asian and Pacific Islander	12,534,653	4.43%
Hispanic	36,613,953	12.94%
	$\Sigma = 283,021,407$	$\Sigma = 100\%$

b. See previous table.

c. First determine the weighted poverty rate for each race, which is the poverty rate for each race times the percentage of the U.S. population represented by each race. The overall U.S. poverty rate is 11.57%. See the following table.

Race	Poverty Rate	Population Weight	Weighted Poverty Rate
White (Non-Hispanic)	7.6%	0.6889	5.24%
Black	22.6%	0.1260	2.85%
American Indian and Alaska Native	22.5%	0.0114	0.26%
Asian and Pacific Islander	10.1%	0.0443	0.45%
Hispanic	21.5%	0.1294	2.78%
			$\Sigma = 11.57\%$

d. The black poverty rate has more impact on the U.S. poverty rate. We can tell this by looking at the weighted poverty rate values. The black weighted poverty rate is larger because of the greater number of individuals of that race.

e. The total number in poverty from all races is 32,742 thousand people. When this number is divided by the U.S. population the result is 11.57%. This percentage is the overall U.S. poverty rate, and matches the rate that we computed in part c).

7. Finding 5-number summaries

   a. The minimum is 26 and the maximum is 69. Because there is an even number of numbers, the median is the mean of the two middle numbers: $(44+50)/2 = 47$. Q1 is the median of all numbers *less than* 47, so it is the mean of 32 and 40: $Q1 = (32+40)/2 = 36$. Q3 is the median of all numbers *greater than* 47, so it is the mean of 59 and 60: $Q3 = (59+60)/2 = 59.5$.

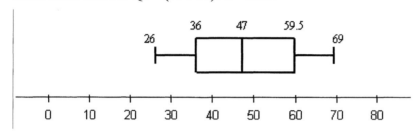

   b. The minimum is 15 and the maximum is 69. Because there is an odd number of numbers, the median is the middle number or 30. There are four numbers less than the median, so Q1 is the mean of the middle two numbers: $Q1 = (15+20)/2 = 17.5$. There are four numbers greater than the median, so Q3 is the mean of the middle two numbers: $Q3 = (60+62)/2 = 61$.

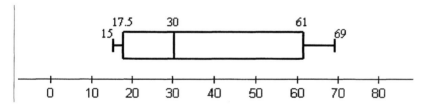

   c. The minimum is 3.4 and the maximum is 5.2. Because there is an even number of numbers in the data set, the median is the mean of the two middle numbers: median $= (4.0+4.0)/2 = 4.0$. Q1 is the median of the first five numbers: $Q1 = 3.8$.

Q3 is the median of the last five numbers: Q3 = 4.6.

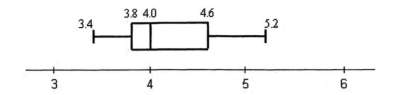

d. The minimum is 100 and the maximum is 160. There is an odd number of numbers, so the median is the middle (6th) number or 150. Q1 is the median of the first five numbers: Q1 = 120. Q3 is the median of the last five numbers: Q3 = 160.

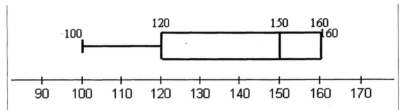

9. Water use

   a. The first quartile in Louisiana uses 129 gallons per person per day.

   b. In Maryland, which has a median of 101 gallons per person per day.

   c. Both Louisiana and Ohio have equally small ranges (spreads). The range for each is 267 gallons per person per day.

   d. County water use is positively skewed (longest tail to the right). The Bowley skews for Maryland, Louisiana and Ohio are given next:

   Maryland: $\text{skew} = \dfrac{122 - 2(101) + 86}{122 - 86} = 0.17$

   Louisiana: $\text{skew} = \dfrac{192 - 2(150) + 129}{192 - 129} = 0.33$

   Ohio: $\text{skew} = \dfrac{136 - 2(110) + 92}{136 - 92} = 0.18$

11. Soil density

   a. To find the 5-number summary by hand, first sort the data in ascending order. Otherwise, input each data set into a technology device, and find the one-variable descriptive statistics.

	Minimum	Q1	Median	Q3	Maximum
Before	0.10	0.145	0.20	0.48	0.81
After	0.15	0.56	0.77	0.96	1.13

b.

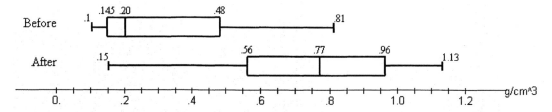

c. The 5 numbers in the boxplot for the post-logging data all lie to the right of those for the pre-logging data. Thus in general, the soil is more compact or denser (greater values in grams per cubic centimeter).

d. The range in the post-logging data ($1.13 - 0.15 = 0.98 \text{ g/cm}^3$) is greater than in the pre-logging data ($0.81 - 0.1 = 0.71 \text{ g/cm}^3$), thus the variation in the post-logging data is greater.

13. Frequency histogram

a. The size of the data set is $n = 20$

b.

Bin (inches)	Frequency	Relative Frequency
20–25	2	10%
25–30	4	20%
30–35	5	25%
35–40	6	30%
40–45	2	10%
45–50	1	5%

c. The modal bin is 35–40 inches. This is the bin with the greatest frequency.

d. In a data set of size $n = 20$, Q1 is the median of the 5th and 6th numbers; in this exercise, Q1 must lie between 25 and 30 inches. Using similar reasoning, the median must lie between 30 and 35 inches, and Q3 between 35 and 40 inches. Note that by examining the relative frequencies, the same conclusions can be drawn.

15. Four histograms

a. Data set #1 is most symmetrical

b. Data set #2 has negative skew (tail to the left).

c. Data sets #3 and #4 have positive skew (tail to the right).

d. Data set #2, because the mean lies closer to the extreme values in the tail, than does the median.

e. Data set #1, because it is symmetrical.

17. Silver fir

a. The mean is $\bar{x} = 15.25$ in. The 5-number summary is: min = 7 in., Q1 = 12 in., median = 15 in., Q3 = 18.5 in., and max = 27 in. The range is 20 in.

b. *Bins and frequencies may vary*

Diameter (inches)	Frequency
7–10	6
10–13	10
13–16	15
16–19	8
19–22	8
22–25	4
25–28	1

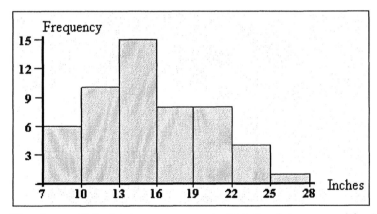

c. The shape of the histogram suggests that the data are positively skewed, although just slightly. The Bowley skew value (0.077) indicates that the middle 50% of the data have positive skew. The histogram shape and Bowley skew value seem to agree.

d. The modal bin is 13-16 inches. This is the most frequently occurring interval of Pacific silver fir diameters.

19. Sampling Atlantic surf clams

   a. Population: All surf clams living in Atlantic waters along the eastern U.S. Sample: The 80 surf clams collected.

   b. 10.2 cm is a statistic, because it refers to the sample.

   c. No, they can only estimate the population mean with the sample mean.

   d. *Answers will vary.* Example: Researchers might only sample shoreline with public access, avoiding areas where access requires permission from land owners. In public areas, it is possible that clam digging is more prevalent; this could result in fewer large clams in the sample. The sample mean, in this case, would be smaller than the population mean.

21. Sampling light bulbs

   a.

Lumens	Watts	Efficiency
210	25	8.4
220	25	8.8
820	60	13.7
830	60	13.8
1,170	75	15.6

Lumens	Watts	Efficiency
1,200	75	16.0
1,600	100	16.0
1,500	100	15.0
1,750	120	14.6
1,800	120	15.0

   b. The sample mean is 13.69 lumens/watt, so we would estimate that the mean energy efficiency of all bulbs in the store (the population) is also 13.69 lumens/watt.

   c. 60% of the sample bulbs have energy efficiencies over 14.0 lumens/watt, so we estimate that the same percentage of bulbs in the store have similar efficiencies.

   d. There are 2 bulbs each with wattages of 25, 60, 75, 100, and 120. It appears that the sample was a systematic sample instead of simple random sample.

Chapter 12

1. Means and standard deviation

   Both sets have means of 25. Set 1 has the greater standard deviation, because the data have, on average, a greater distance from the mean.

3. Temperatures

   Daily temperatures in Portland have a smaller standard deviation because Portland has smaller temperature fluctuations compared to Chicago. Portland temperatures are moderated by the Pacific Ocean. Chicago, not close to an ocean, has extremely hot summers and extremely cold winters.

5. Comparing two histograms

   a. The means and medians for both data sets will be approximately the same because both histograms are symmetrical.

   b. The histogram for data set #1 has more values farther from the center as compared to data set #2. Therefore, the standard deviation of data set #1 is greater than that of data set #2.

7. Potassium concentrations

   a. $\bar{x} = 0.84$ mg/L

x	$\bar{x}$	$x - \bar{x}$	$(x - \bar{x})^2$
1.5 mg/L	0.8417 mg/L	0.6583 mg/L	0.4334 mg^2/L^2
1.8 mg/L	0.8417 mg/L	0.9583 mg/L	0.9183 mg^2/L^2
1.6 mg/L	0.8417 mg/L	0.7583 mg/L	0.5750 mg^2/L^2
1.3 mg/L	0.8417 mg/L	0.4583 mg/L	0.2100 mg^2/L^2
0.7 mg/L	0.8417 mg/L	-0.1417 mg/L	0.0201 mg^2/L^2
0.7 mg/L	0.8417 mg/L	-0.1417 mg/L	0.0201 mg^2/L^2
0.6 mg/L	0.8417 mg/L	-0.2417 mg/L	0.0584 mg^2/L^2
0.6 mg/L	0.8417 mg/L	-0.2417 mg/L	0.0584 mg^2/L^2
0.4 mg/L	0.8417 mg/L	-0.4417 mg/L	0.1951 mg^2/L^2
0.4 mg/L	0.8417 mg/L	-0.4417 mg/L	0.1951 mg^2/L^2
0.4 mg/L	0.8417 mg/L	-0.4417 mg/L	0.1951 mg^2/L^2
0.1 mg/L	0.8417 mg/L	-0.7417 mg/L	0.5501 mg^2/L^2
		$\Sigma(x - \bar{x}) = 0.0004$ mg/L ≈ 0 mg/L	$\Sigma(x - \bar{x})^2 = 3.4291$ mg^2/L^2

The sample standard deviation is the square root of sum of the squared deviations divided by $n-1$: $s = \sqrt{\dfrac{3.4291 \text{ mg}^2/\text{L}^2}{11}} \approx 0.56 \text{ mg/L}$

b. The measurement of 1.8 mg/L lies furthest from the mean, because it has the largest deviation from the mean in absolute value: $|x - \bar{x}| = 0.9583$. The two measurements of 0.7 mg/L lie closest to the mean. The absolute value of their deviation from the mean is $|x - \bar{x}| = 0.1417$.

9. Stream velocities

a. $z = \dfrac{x - \bar{x}}{s} = \dfrac{1.8 \text{ mph} - 2.10 \text{ mph}}{0.60 \text{ mph}} = -0.5$

b. $z = \dfrac{x - \bar{x}}{s} = \dfrac{5.0 \text{ mph} - 2.10 \text{ mph}}{0.60 \text{ mph}} \approx 4.8$

c. $z = 0$

d. $z = -2.6$ (The z-score *is* the number of standard deviations above or below the mean.)

11. Temperatures

a. Substitute $\bar{x}, s,$ and z into the z-score formula and then solve for x. The solution is

$x = 18.64 \,^{\circ}\text{C} + 0.75(0.08 \,^{\circ}\text{C}) = 18.7 \,^{\circ}\text{C}$

b. Substitute $\bar{x}, s,$ and z into the z-score formula and then solve for x. The solution is

$x = 18.64 \,^{\circ}\text{C} - 3(0.08 \,^{\circ}\text{C}) = 18.4 \,^{\circ}\text{C}$

c. The mean, which is $18.64 \,^{\circ}\text{C}$.

d. Temperatures greater than the mean have positive z-scores, while those less than the mean have negative z-scores.

13. Iowa

Nuclear: $z = \dfrac{3,000 \text{ GWh} - 17,832.8 \text{ GWh}}{15,714.6 \text{ GWh}} = -0.94$

Hydro: $z = \dfrac{13 \text{ GWh} - 6,225.6 \text{ GWh}}{14,272.9 \text{ GWh}} = -0.44$

As compared to the means of the respective industries, Iowa's nuclear energy production is smaller than its hydroenergy production.

15. Nuclear energy

Pennsylvania produced 57,800 GWh, the second greatest amount of nuclear energy in 1990 among the U.S. states. The z-score for this amount is

$z = \dfrac{57,800 \text{ GWh} - 17,832.8 \text{ GWh}}{15,714.6 \text{ GWh}} = 2.54$, below the $z = 3$ cutoff level for outliers.

17. Which are true?

Statements b) and d) are true. If at least 75% of the data are within two standard deviations of the mean, then no more than 25% are farther than two standard deviations from the mean.

19. Using the Empirical Rule

a. The mean is about 21. About 99.7% of the area under the bell curve lies with three standard deviations of the mean. To estimate the standard deviation, estimate the distance from the end of either tail to the mean. Divide that distance by 3 to get the standard deviation. You can also estimate the horizontal distance from the mean to either inflection point—that distance will equal the standard deviation. The standard deviation is about 6.

b. The mean is about 140. One method to estimate the standard deviation is to take the distance from the end of either tail to the mean (120) and divide 3. The result is 40. Alternatively, make an estimate of the original data set and then find the standard deviation. A typical way to do this involves assigning frequencies to each histogram bar in proportion to the bar heights and letting each bar be represented by the bin midpoint. For example, the bar heights from left to right are 1, 3, 5, 7, 7, 5, 3, and 1. The bin midpoints from left to right are 35, 65, 95, 125, 155, 185, 215, and 245. Thus the data set consists of one value of 35, three values of 65, five values of 95, and so on. The standard deviation of this data set is 50.5. Thus a reasonable estimate of the standard deviation for this histogram is from 40 to 50.

21. Motor oil recycling rates

a. $\bar{x} = 1.816$ pounds per person, $s = 1.196$ pounds per person.

b. *Bins may vary.*

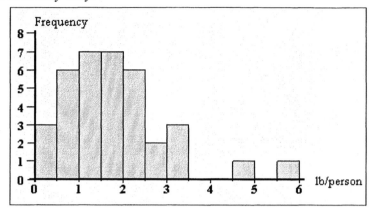

c. 34 out of the 36 counties (94.4%) have oil recycling values within two standard deviations of the mean (from -0.576 to 4.208 pounds per person). This agrees with Chebychev's Rule, which states that at least 75% of the data will fall within two standard deviations of the mean.

d. Asotin county recycles 5.63 pounds per person, which corresponds to a z-score of $z = \dfrac{5.63\,\text{lb/person} - 1.816\,\text{lb/person}}{1.196\,\text{lb/person}} = 3.2$. Because the rate is more than three standard deviations above the mean, Asotin is an outlier.

e. It's possible that those counties have mandatory recycling, very few industries, more recycling centers, or stronger recycling education.

23. A study of 58 small businesses

For all questions, it may be helpful to look at a bell-curve marked with the garbage weights that lie one, two and three standard deviations from the mean (see the following diagram). The percentages are based on the Empirical rule (see Figure 12-11 in the text).

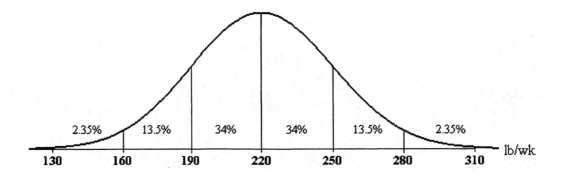

a. 190 lb is one standard deviation below the mean, while 250 lb is one standard deviation above the mean. By the Empirical rule, 68% of a normally distributed data set will lie within one standard deviation of the mean. Thus 68% of the businesses discarded between 190 and 250 pounds/week.

b. 160 lb is two standard deviations below the mean, while 280 lb is two standard deviations above the mean. By the Empirical rule, 95% of a normally distributed data set will lie within two standard deviations of the mean. Thus $0.95(58) \approx 55$ businesses discarded between 160 and 280 pounds/week.

c. 5% of a normally distributed data set has lies more than two standard deviations above or below the mean. Because of symmetry, half of that, or 2.5% of the data, lie two or more standard deviations above the mean. Thus $0.025(58) \approx 1$ business discarded more than 280 pounds/week.

d. 50% + 34% = 84% of the businesses discard cardboard below 250 pounds/week.

Chapter 13

1. Empirical Rule

   a. By the Empirical Rule, the area under the standard normal curve between $z = -3$ and $z = 3$ is 0.997. So half that area is $0.5(0.997) = 0.4985 = 49.85\%$.

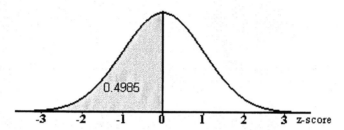

   b. By the Empirical Rule, the area under the standard normal curve between $z = -1$ and $z = 1$ is 0.68. So the area between $z = 0$ and $z = 1$ is 0.34 (see the following figure). The area to the right of $z = 1$ is $0.5 - 0.34 = 0.16 = 16\%$.

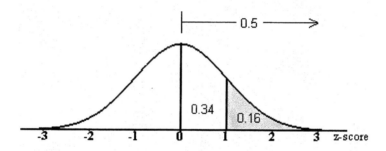

   c. By the Empirical Rule, the area under the standard normal curve between $z = -2$ and $z = 2$ is 0.95, so the area between $z = -2$ and $z = 0$ is 0.475 (see the following figure). Also by the Empirical Rule, the area under the standard normal curve between $z = -3$ and $z = 3$ is 0.997, so the area between $z = 0$ and $z = 3$ is 0.4985. The total area between $z = -2$ and $z = 3$ is $0.475 + 0.4985 = 0.9735 = 97.35\%$.

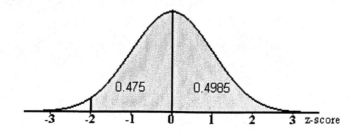

3. Standard normal curve

    a. In Table 13-1, look in the row labeled **1.2** and in the column labeled **.05**. The area is 0.3944.

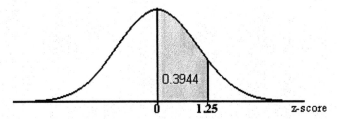

    b. In Table 13-1, look in the row labeled **2.2** and in the column labeled **.00**. The area is 0.4861.

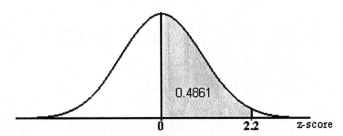

    c. The area between $z = -2.28$ and $z = 0$ is the same as the area between $z = 0$ and $z = 2.28$. In Table 13-1, look in the row labeled **2.2** and in the column labeled **.08**. The area is 0.4887.

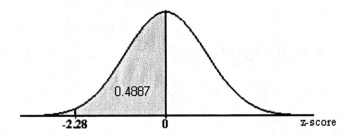

5. Standard normal curve

   a. In Table 13-1, look in the row labeled **2.6** and in the column labeled **.01**. The area is 0.4955. The area to the right of $z = 2.61$ is $0.5 - 0.4955 = 0.0045$

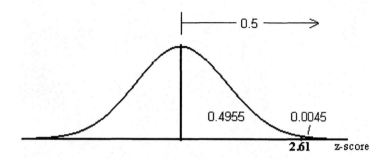

   b. The area between $z = -2.2$ and $z = 0$ is the same as the area between $z = 0$ and $z = 2.2$. In Table 13-1, look in the row labeled **2.2** and in the column labeled **.00**. The area is 0.4861 (see the following figure). To find the area between $z = 0$ and $z = 1.62$, look in the row labeled **1.6** and in the column labeled **.02**. The area is 0.4474. The total area equals $0.4861 + 0.4474 = 0.9335$.

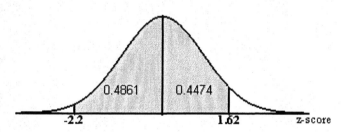

   c. First find the area between $z = 0$ and $z = 0.6$ by looking in the row labeled **0.6** and in the column labeled **.00**. The area is 0.2257 (see the following figure). Table 13-1 also indicates that the area under the curve between $z = 0$ and $z = 2.8$ is 0.4974. Subtracting these areas results in the solution: $0.4974 - 0.2257 = 0.2717$.

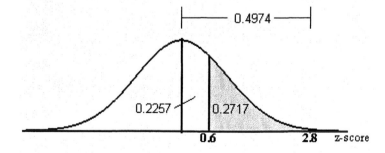

85

7. Estimate the z-score

    a. This is an inverse problem in which the area is given and we need to find the correct right-hand bound (z-score) denoted as $z = c$. The left-hand bound is $z = 0$, so we can use Table 13-1 directly. Search the table for an area of 0.4671. The area is found in row **1.8** and column **.04**, thus $c = 1.84$.

    b. In the problem we see that the value of c must be negative, as $z = c$ lies to the left of $z = 0$. Search the table for an area of 0.375. The closest area is 0.3749, found in row **1.1** and column **.05**. Thus the area between $z = 0$ and $z = 1.15$ is about 0.375. Because the value of c is negative, $c = -1.15$.

    c. The area to the right of $z = 0$ is 0.5, so the area between $z = c$ and $z = 0$ is $0.968 - 0.5 = 0.468$. The closest area in the table is 0.4678, which is listed in row **1.8** and **.05**. In this problem we know that c is a negative z-score, thus $c = -1.85$.

9. SUVs

The z-score for 15.8 mpg is $z = \dfrac{15.8 \text{ mpg} - 16.6 \text{ mpg}}{3.3 \text{ mpg}} = -0.24$. The z-score for 21.4 mpg is $z = \dfrac{21.4 \text{ mpg} - 16.6 \text{ mpg}}{3.3 \text{ mpg}} = 1.45$. The area under the standard normal curve between these two z-scores can be found by adding the area between $z = -0.24$ and $z = 0$ (which is 0.0948) to the area between $z = 0$ and $z = 1.45$ (which is 0.4265).

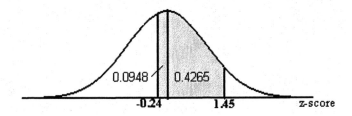

The result is $0.0948 + 0.4265 = 0.5213 = 52.13\%$. Approximately 52% of the SUVs have mileage ratings between 15.8 and 21.4 mpg.

11. pH of water samples

    a. The z-scores for the pH values of 5.5 and 5.8 are $z = 0$ and $z = 0.75$, respectively. The area under the standard normal curve between these two z-scores is 0.2734. The percentage of water samples is 27.34%.

b. The z-scores for the pH values of 4.9 and 5.5 are $z = -1.5$ and $z = 0$, respectively. The area under the standard normal curve between these two z-scores is 0.4332. The percentage of water samples is 43.32%.

c. The z-score for a pH of 5.7 is $z = 0.5$. The area we seek under the standard normal curve is shown next. The area is $0.5 - 0.1915 = 0.3085$.

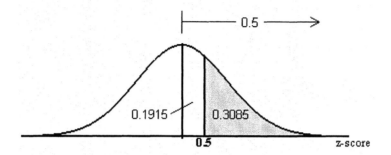

We've found that 30.85% of the samples have pH values greater than 5.7, which is about 20 samples.

d. This is an inverse problem in which we are given the area and need to find the z-score $z = c$. See the following figure.

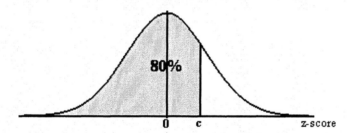

We seek a z-score $z = c$ such that the area between $z = 0$ and $z = c$ is 30% or 0.3000. Using Table 13-1, we find that $c = 0.84$ is about the correct number.

Now we need to find the pH value that is associated with a z-score of $z = 0.84$. To do this, we solve the equation $0.84 = \dfrac{x - 5.5}{0.4}$ for x. The answer is $x = 5.836$, indicating a pH of about 5.8.

13. Moisture in soils

a. For a water content of 0.34 m^3/m^3, the z-score is $z = 1.75$. The area that lies to the left of this z-score is $0.5 + 0.4599 = 0.9599$ (see the following figure). The percentage of soil cores is about 96%.

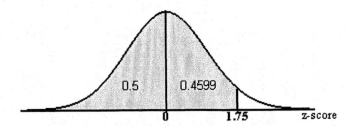

b. The soil water contents of 0.21 m³/m³ and 0.35 m³/m³ have z-scores equal to $z = -1.5$ and $z = 2$, respectively. The area under the standard normal curve is $0.4332 + 0.4772 = 0.9104$ (see the following figure). The percentage of soil cores is about 91%.

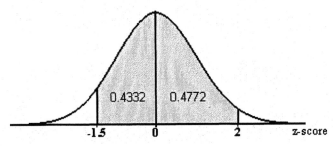

c. We are given an area of 40% and need to find the negative z-score $z = c$. See the following figure.

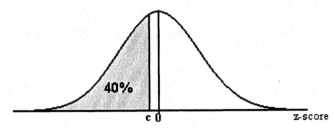

The area between $z = c$ and $z = 0$ is $10\% = 0.1000$. In Table 13-1, we find that the *positive* z-score of $z = 0.25$ is such that the area between $z = 0$ and itself is about 0.1. Because c must be negative, its value is $c = -0.25$.

Now take the z-score of $z = -0.25$ and find the soil moisture content, given by x. Solve the equation $-0.25 = \dfrac{x - 0.27}{0.04}$ to get $x = 0.26$ m³/m³.

15. Coal-fired power plants

   a. *Histograms will vary depending on binning strategy.*

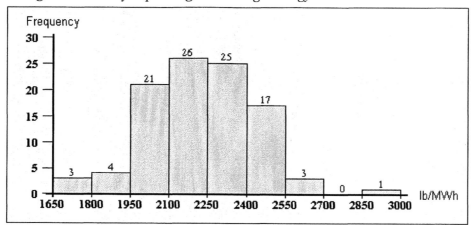

   b. The histogram has a decent bell shape, centered around 2,250 lb/MWh. The power plant that produces 2,982 lb/MWh appears to be an outlier. The z-score for this power plant is $z = \dfrac{2{,}982 \text{ lb/MWh} - 2{,}223.1 \text{ lb/MWh}}{211.3 \text{ lb/MWh}} \approx 3.6$, confirming that it is an outlier.

   c. By Table 13-1, the percentage of the area below the standard normal curve from $z = 0$ to $z = 1.5$ is 43.32%. Thus the area to the right of $z = 1.5$ represents $50\% - 43.32\% = 6.68\%$ of the area. In theory, about 7% of the data should have z-scores greater than 1.5.

   d. 1.5 standard deviations above the mean correspond to a CO_2 output rate of $2{,}223.1 + 1.5(211.3) = 2{,}540.05$ lb/MWh. Only 4 of the 100 power pants (4%) have CO_2 outputs above this level (which is reasonably close to 7%).

   e. The Marion plant's z-score is $z = \dfrac{2{,}590 \text{ lb/MWh} - 2{,}223.1 \text{ lb/MWh}}{211.3 \text{ lb/MWh}} \approx 1.74$. The region under the standard normal curve that is left of this z-score is $0.5 + 0.4591 = 0.9591$. In theory, about 96% of the power plants should have CO_2 production rates smaller than that of the Marion plant. The actual percentage is 98%.

17. Finding 95% confidence intervals for means

   a. $E = 2.262 \dfrac{2.2 \text{ lb}}{\sqrt{10}} \approx 1.6$ lb. The 95% confidence interval is (18.9 lb, 22.1 lb) or 20.5 ± 1.6 lb.

b. $E = 2.069 \dfrac{5 \text{ km}}{\sqrt{24}} \approx 2 \text{ km}$. The 95% confidence interval is (43 km, 47 km) or

45 ± 2 km.

c. $E = 1.960 \dfrac{0.15 \text{ cfs}}{\sqrt{50}} \approx 0.04$ cfs. The 95% confidence interval is (3.15 cfs, 3.23 cfs)

or 3.19 ± 0.04 cfs.

19. Lead and bridge workers

   a. $E = 1.960 \dfrac{16.1 \; \mu\text{g/dL}}{\sqrt{373}} \approx 1.6 \; \mu\text{g/dL}$. The 95% confidence interval is

   $27.2 \pm 1.6 \; \mu\text{g/dL}$ or $(25.6 \; \mu\text{g/dL}, 28.8 \; \mu\text{g/dL})$.

   b. We are at least 95% certain that the mean blood lead level for all bridge workers is above $25 \; \mu\text{g/dL}$, because the lower boundary of the 95% confidence interval is $25.6 \; \mu\text{g/dL}$. Bridge workers should modify their exposure to lead to lower their mean blood level.

21. Sea stars

   a. $E = 1.960 \dfrac{2.00 \text{ cm}}{\sqrt{67}} \approx 0.48$ cm

   b. In the denominator of the E formula, an extra factor of 4 inside the square root is equivalent to an extra factor of 2 outside the square root. An extra 2 in the denominator would reduce the margin of error by a factor of 2 (cut in half).

   c. We know that the sample size would be much larger than 67, so we know that t will equal 1.960. To find the value of n, we solve the inequality

   $1.960 \dfrac{2.00 \text{ cm}}{\sqrt{n}} < 0.10$ cm. First multiply each side of the equation by $\sqrt{n}$, and

   divide each side by 0.10 cm to get: $\dfrac{1.960(2.00 \text{ cm})}{0.10 \text{ cm}} < \sqrt{n}$. Simplifying the left side

   results in $39.2 < \sqrt{n}$. Now square each side to get $1{,}536.64 < n$. The sample size must be at least $n = 1{,}537$ sea stars.

23. Finding 95% confidence intervals for proportions

a. $40(0.74) = 29.6 \geq 5$ ✓ and $40(1-0.74) = 10.4 \geq 5$ ✓

$$E = 1.960\sqrt{\frac{0.74(1-0.74)}{40}} \approx 0.14 = 14\%$$

The 95% confidence interval is $(60\%, 88\%)$ or $74\% \pm 14\%$.

b. $15(0.38) = 5.7 \geq 5$ ✓ and $15(1-0.38) = 9.3 \geq 5$ ✓

$$E = 2.145\sqrt{\frac{0.38(1-0.38)}{15}} = 0.27 = 27\%$$

The 95% confidence interval is $(11\%, 65\%)$ or $38\% \pm 27\%$.

c. $1,000(0.354) = 354 \geq 5$ ✓ and $1,000(1-0.354) = 646 \geq 5$ ✓

$$E = 1.960\sqrt{\frac{0.354(1-0.354)}{1,000}} = 0.0296 \approx 3.0\%$$

The 95% confidence interval is $(32.4\%, 38.4\%)$ or $35.4\% \pm 3.0\%$.

25. StarLink corn

a. This large of a sample should make the margin of error quite small.

$$E = 1.960\sqrt{\frac{0.09(1-0.09)}{110,000}} \approx 0.002 = 0.2\%$$

b. No, the margin of error is much smaller than the precision implied with the statement "about 9%."

27. Gallup poll

a. $526(0.55) = 289.3 \geq 5$ ✓ and $526(1-0.55) = 236.7 \geq 5$ ✓

b. $E = 1.960\sqrt{\frac{0.55(1-0.55)}{526}} \approx 0.04 = 4\%$, so the 95% confidence interval is $55\% \pm 4\%$. We can be 95% confident that the true population percentage is between 51% and 59%.

c. To find the minimum sample size, solve the inequality

$$1.960\sqrt{\frac{0.55(1-0.55)}{n}} < 0.03.$$

First divide by 1.960 and square each side: $\dfrac{0.55(1-0.55)}{n} < \left(\dfrac{0.03}{1.960}\right)^2$.

Next multiply each side by n and by $\left(\dfrac{0.03}{1.960}\right)^2$ to get $\dfrac{0.55(1-0.55)}{\left(\dfrac{0.03}{1.960}\right)^2} < n$.

Simplify to find that $n > 1{,}056.44$. We would have to sample a minimum of 1,057 people to ensure that $E = 3\%$.